£10-00

Lecture Notes in Mathematics

Edited by A. Dold and B. Eckmann

D1423614

958

F. Rudolf Beyl
Jürgen Tappe

Group Extensions, Representations, and the Schur Multiplicator

Springer-Verlag
Berlin Heidelberg New York 1982

Authors
F. Rudolf Beyl
Mathematisches Institut der Universität
Im Neuenheimer Feld 288, 6900 Heidelberg, Germany

Jürgen Tappe
Lehrstuhl für Mathematik
Rhein.-Westf. Technische Hochschule Aachen
Templergraben 55, 5100 Aachen, Germany

AMS Subject Classifications (1980): 20C25, 20E22, 20J05, 20C20, 20E10, 20J06

ISBN 3-540-11954-X Springer-Verlag Berlin Heidelberg New York
ISBN 0-387-11954-X Springer-Verlag New York Heidelberg Berlin

Printing and binding: Beltz Offsetdruck, Hemsbach/Bergstr.
2146/3140-543210

TABLE OF CONTENTS

INTRODUCTION

The aim of these notes is a unified treatment of various group-theoretic topics for which, as it turns out, the Schur multiplicator is the key. At the beginning of this century, classical projective geometry was at its peak, while representation theory was growing in the hands of Frobenius and Burnside. In this climate our subject started with the two important papers of Schur [1],[2] on the projective representations of finite groups.

But it was only in the light of the much more recent (co)homology theory of groups that the true nature of Schur's "Multiplicator" and its impact on group theory was fully realized; the papers by GREEN [1], YAMAZAKI [1], STALLINGS [1], and STAMMBACH [1] have been most influential in this regard.

The first chapter provides the setting for these notes. We start out with the concepts of group extension (handled in terms of diagrams) and Schur multiplicator (here defined by the Schur-Hopf Formula) to obtain a group-theoretic version of the Universal Coefficient Theorem. All these concepts and the Ganea map have a homological flavor, but are here developed in a rather elementary group-theoretic fashion; the (co)homology theory of groups is not a prerequisite for reading most of these notes. The first chapter also includes a full translation from our approach to the usual group (co)homology for the reader's convenience. (We feel that our presentation is very suited for the applications to follow, but this view is to some extent a matter of taste.)

In the second chapter we consider projective representations, which can be regarded as homomorphisms into projective groups. Schur showed that the projective representations of the finite group Q over the complex field $\mathbb{C}$ can be described in terms of the (linear) representations of certain central extensions by Q, and thus invented the "Darstellungsgruppen von Q" or, in English, the representation groups of Q. In the course of this chapter, several variations of Schur's theme are discussed. (The problem is that of lifting homomorphisms where the projective representations are replaced by more general homomorphisms.) In most of the literature on Schur's theory, representation groups etc. are defined only for finite groups. In these notes this attitude is seen to be unnecessarily restrictive, provided one carefully distinguishes between M(Q), as the common kernel of all representation groups, and the competing condidate $H^2(Q,\mathbb{C}^*) \cong \mathrm{Hom}(M(Q),\mathbb{C}^*)$. We find that representation groups of arbitrary groups are important in many parts of group theory beyond the original aspects of representation theory. For example, the final section of Chapter II gives a comprehensive treatment of the covering theory of perfect groups. This theory is related e.g. to finite simple groups, but has recently gained particular attention for its relevance to Milnor's K_2 functor - where Q is an infinite matrix group.

In the third chapter we study the notion of isoclinism, which P. Hall introduced in his Göttingen lectures on the classification of groups. In spite of World War II having begun, summaries of Hall's lectures were published in "Crelle's Journal", cf. P. HALL [1],[2],[3],[4]; it may be due to these circumstances that Hall did not publish further details. The notion of isoclinism is extended to central extensions; this step is more or less technical, but provides for clarity and enables us to apply the machinery of Chapter I. The first major result of this chapter is a description of isoclinism classes in terms of the subgroups of the Schur multiplicator. We then study a refinement of the isoclinism concept and prove formulae of P. Hall [3]. Our treatment in terms of central extensions differs from Hall's, which employs free presentations. In any case, Chapter III brings out some connections between both views.

In the final section of Chapter III, we work out implications of the isoclinism relation for the ordinary and the modular representations of finite groups.

The last chapter contains further group-theoretic applications of the Schur multiplicator, in some aspects it supplements STAMMBACH [3]. We first resume the question of group deficiency, a concept grown out of the desire to present a group with as few relators as possible. Our emphasis is on worked-out examples and their interpretation. Among other results, we give rather elementary treatments of (i) Swan's examples of finite groups with trivial multiplicator and large deficiency; (ii) an interesting representation group of the non-abelian group of order p^3 and exponent p, for p an odd prime; (iii) metacyclic groups and their multiplicators. The next topic is a group invariant $Z^*(G)$, the central subgroup that measures how much G deviates from being a group of inner automorphisms. (This concept is unrelated to Glauberman's Z^*, any serious confusion seems to be unlikely.) We then obtain rather explicit results on the question whether a central group extension lies in a given variety of exponent zero. These sections again show the importance of Schur's "hinreichend ergänzte Gruppen", i.e.

central extensions having the lifting property for complex projective representations; they are called "generalized representation groups" in these notes. The chapter ends with a development of isologism, a related concept of P. Hall, in analogy with our treatment of isoclinism in Chapter III. The reader will now be prepared to study LEEDHAM-GREEN/McKAY [1] and other papers on varietal cohomology.

These notes partly present results from our "Habilitationsleistungen" at "Ruprecht-Karls-Universität Heidelberg" and "Rheinisch-Westfälische Technische Hochschule Aachen", respectively. We thankfully acknowledge the support we received from our institutions, as well as partial support from the "Deutsche Forschungsgemeinschaft (DFG)", the "Forschungsinstitut für Mathematik (ETH Zürich)", and the "Gesellschaft von Freunden der Aachener Hochschule (FAHO)".

We remember with pleasure that we greatly profited from the feedback various seminar audiences gave us, in particular from discussions with the late R. Baer, with P. Hilton, C.R. Leedham-Green, R. Laue, J. Neubüser, J. Ritter, U. Stammbach, R. Strebel, and J. Wiegold.

CHAPTER I. GROUP EXTENSIONS WITH ABELIAN KERNEL

The core of this chapter consists of Sections 3 and 4.

1. The Calculus of Induced Extensions

This section is preparatory. We introduce forward and backward induced group extensions and discuss the relationship between extensions and factor systems. An extension of the group N by the group Q is a short exact sequence

(1.1) $e = (\varkappa, \pi) : N \rightarrowtail G \twoheadrightarrow Q$

or, equivalently, an exact sequence

$$e : 0 \longrightarrow N \xrightarrow{\varkappa} G \xrightarrow{\pi} Q \longrightarrow 0$$

of groups. The arrows $\rightarrowtail$ and $\twoheadrightarrow$ denote injective and surjective homomorphisms (monomorphisms and epimorphisms), respectively, and $0 = \{1\}$ stands for the group of one element. (This terminology accords with category theory and is easy to work with.)

In most of the cases treated here, N will be abelian. Then the extension e gives rise to a Q-module structure on N, which is well-defined by ${}^{q}n = \varkappa^{-1}(g\cdot\varkappa n\cdot g^{-1})$, where $q \in Q$, $n \in N$ and $g \in G$ is any element with $\pi g = q$. Now let $(N, \varphi\colon Q \to \mathrm{Aut}(N))$ be a Q-module. We call e a Q-extension of (N,φ) if e is an extension of N by Q which induces the given Q-module structure on N by the method above. Two extensions e_1 and e_2 of N by Q are called congruent, if there exists an isomorphism $\beta\colon G_1 \to G_2$ such that the following diagram is commutative; we then write $e_1 \equiv e_2$.

$$
(1.2)\qquad
\begin{array}{ccccccc}
e_1 : & N & \rightarrowtail & G_1 & \twoheadrightarrow & Q \\
 & \| & & \downarrow{\scriptstyle \beta} & & \| \\
e_2 : & N & \rightarrowtail & G_2 & \twoheadrightarrow & Q
\end{array}
$$

Congruence is an equivalence relation. When N is abelian, congruent extensions define the same Q-module structure on N. Let $\mathrm{Opext}(Q,N,\varphi)$ denote the set of congruence classes $[e]$ of Q-extensions of (N,φ) . The existence of the extension

$$
(1.3)\qquad e_o : N \overset{\varkappa_o}{\rightarrowtail} N \rtimes Q \overset{\pi_o}{\twoheadrightarrow} Q
$$

implies that $\mathrm{Opext}(Q,N,\varphi)$ is not empty. - Here $N \rtimes Q$ is the <u>semidirect product</u> of Q by the Q-module N with the multiplication formula $(n,q)\cdot(n_1,q_1) = (n\cdot {}^q n_1, q\cdot q_1)$. The maps $\varkappa_o$ and π_o are defined by $\varkappa_o(n) = (n,1)$ and $\pi_o(n,q) = q$. - The extension e as in (1.1) is called <u>split</u> if there is a homomorphism $\sigma: Q \to G$ with $\pi\sigma = 1_Q$. The Q-extension e of (N,φ) is split precisely when it is congruent to e_o .

A <u>morphism</u> $(\alpha,\beta,\gamma): e_1 \to e_2$ <u>of</u> <u>extensions</u> is a commutative diagram

$$
(1.4)\qquad
\begin{array}{ccccccc}
e_1 : & N_1 & \rightarrowtail & G_1 & \twoheadrightarrow & Q_1 \\
 & \downarrow{\scriptstyle \alpha} & & \downarrow{\scriptstyle \beta} & & \downarrow{\scriptstyle \gamma} \\
e_2 : & N_2 & \rightarrowtail & G_2 & \twoheadrightarrow & Q_2
\end{array}
$$

This is called an <u>isomorphism</u> <u>of</u> <u>extensions</u> if α and γ (hence also β) are group isomorphisms. Now assume N_i abelian and let $\varphi_i: Q_i \to \mathrm{Aut}(N_i)$ be the structual maps defined by e_i , for $i:=1,2$. If a morphism $(\alpha,\beta,\gamma): e_1 \to e_2$ exists, then

$$
(1.5)\qquad \alpha({}^q n) = {}^{\gamma(q)}\alpha(n) \quad \text{for all } n \in N_1 ,\ q \in Q_1
$$

or, equivalently, $\alpha: (N_1,\varphi_1) \to (N_2,\varphi_2\gamma)$ is Q_1-homomorphic.

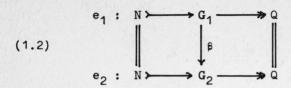

If the extension e as in (1.1) is central, i.e. $\varkappa(N)$ lies in the center $Z(G)$ of G, then the associated map $\varphi: Q \to \mathrm{Aut}\, N$ is trivial. The converse is also true. We write 0 for the trivial homomorphism between any two groups and $\mathrm{Cext}(Q,N)$ for $\mathrm{Opext}(Q,N,\varphi=0)$. If moreover Q is abelian, $\mathrm{Cext}(Q,N)$ contains the set $\mathrm{Ext}(Q;N)$ of congruence classes of abelian extensions (i.e. G abelian) as a subset. This interpretation of $\mathrm{Ext}(Q;N)$ agrees with that of MAC LANE[2; III,§1].

Once we have shown that a certain construction using extensions yields the same for congruent extensions, we allow ourselves to confuse congruence classes with representing extensions in this context.

1.1. Here we shortly review the notion of factor systems, although we try to avoid direct computations with factor systems and cocycles as far as possible. If you are given the extension (1.1) with N abelian and operation $\varphi: Q \to \mathrm{Aut}\, N$, choose a transversal $\{ u_q \in G \mid \pi(u_q) = q \in Q \}$ to $\varkappa N$ in G, i.e. a system of coset representatives. The map $\{q \longmapsto u_q\}: Q \to G$ will not be a homomorphism, in general. The deviation is measured by the associated factor system $f: Q \times Q \to A$, this set map being defined by

$$\varkappa\, f(r,s) = u_r \cdot u_s \cdot u_{r\cdot s}^{-1} \quad \text{for} \quad r,s \in Q .$$

The associative law $(u_r u_s)u_t = u_r(u_s u_t)$ implies the formula

(1.6) $\quad f(r,s)\cdot f(r\cdot s,t) = f(r,s\cdot t) \cdot {}^{r}f(s,t) \qquad r,s,t \in Q .$

If we choose another transversal $\{u'_q\}$, then the new factor system is $f'=f+\delta g: Q \times Q \to A$, where $g = \{q \longmapsto \varkappa^{-1}(u'_q \cdot u_q^{-1})\}: \ Q \to A$ and

(1.7) $\quad (\delta g)(r,s) := g(r)\cdot {}^{r}g(s)\cdot g(r\cdot s)^{-1} \qquad r,s \in Q .$

Next we consider all functions f which satisfy (1.6) and call them factor sets. For every function $g: Q \to A$ we define $\delta g: Q \times Q \to A$ by (1.7); we note that δg is a factor set and call it a principal factor set or transformation set. For computational details regarding this subsection consult MAC LANE [2; IV, §4].

The factor sets form an additive abelian group, with addition defined by pointwise multiplication, and the principal factor sets are a subgroup thereof. If we are given a congruence $e = e_1 \equiv e_2$ as in (1.2), then we may choose the transversals compatible, i.e. $u_q^{(2)} = \beta u_q$, and conclude that these choices yield the same factor system for e_1 and e_2. All told, every congruence class of Q-extensions of (N,φ) determines a unique element in the group $H(Q,N,\varphi) := \{\text{factor sets}\}/\{\text{principal factor sets}\}$.

When choosing the transversal $\{u_q\}$, we are free to require $u_1 = 1$. Then $f(q,1) = f(1,q) = 1$ for all $q \in Q$; we call such a factor set normalized. Condition (1.6) implies $f(q,1) = {}^q f(1,1)$ and $f(1,q) = f(1,1)$ for any factor set; we can replace f by $f - \delta g$, where $g: Q \to A$ is the constant function with value $f(1,1)$; then $f - \delta g$ is normalized and determines the same element in $H(Q,A,\varphi)$ as f does. For any $g: Q \to A$, formula (1.7) gives $g(1) = (\delta g)(1,1)$; hence δg is a normalized factor set exactly if $g(1) = 1$. We conclude that $H(Q,N,\varphi)$ is isomorphic to $\{\text{normalized factor sets}\}/\{\text{normalized principal factor sets}\}$.

1.2. Let an element $X \in H(Q,N,\varphi)$ be given, is there some Q-extension e of (N,φ) yielding this X by the procedure of 1.1? If we specify a factor set f in X, is it possible to obtain f as a factor system for e and a suitable choice of coset representatives? The answer is "Yes" in both cases, as the following construction shows.

Any factor set in X has the form $f + \delta g$ where, without loss of generality, f is a normalized factor set and g is some function $Q \to A$. (The normalization assumption on f is for convenience only, the given product formula also works without it.) Construct a group G with underlying set $N \times Q$ and multiplication

$$(n,q)\cdot(n_1,q_1) = (n\cdot{}^{q}n_1\cdot f(q,q_1),\ q\cdot q_1) .$$

This is indeed a group: associativity follows from (1.6), the neutral element is $(1,1)$, and $(n,q)^{-1} = ({}^{y}(n^{-1})\cdot f(y,q)^{-1},y)$ with $y:=q^{-1}$. Moreover the maps $\varkappa\colon N \to G$ and $\pi\colon G \to Q$, defined by $\varkappa(n) = (n,1)$ and $\pi(n,q) = q$, are homomorphisms such that $0 \longrightarrow N \xrightarrow{\varkappa} G \xrightarrow{\pi} Q \longrightarrow 0$ is a Q-extension of (N,φ). If we choose $\{u'_q = (g(q),q)\}$ as a transversal, then the associated factor system is $f + \delta g$. For computational details consult e.g. NORTHCOTT [1; §10.10].

It was a great stimulus to the cohomology theory of groups, when EILENBERG/MAC LANE [1], [2] discovered that the set of congruence classes of group extensions is isomorphic to the second cohomology group of Q in the Q-module (N,φ). By computing cohomology groups at a particular resolution, viz. the normalized bar resolution in inhomogeneous form, they were led to the functional equations (1.6) and (1.7) and thus found the second cohomology group also given by {normalized factor sets}/{normalized principal factor sets}.

Now the idea of factor systems is near at hand and had already been used by HÖLDER [1] in 1893. Later SCHREIER [1],[2] continued and systematized the treatment of group extensions (also non-abelian groups N allowed) via factor systems. The transition to the second cohomology group via a particular resolution looks artificial, however. In this chapter we try to present a reasonably complete theory of group extensions without resolutions.

1.3 DEFINITION. Given a Q-extension e of the Q-module (N,φ) and a group homomorphism $\gamma\colon Q_1 \to Q$. We define a Q_1-extension $e\gamma$

of $(N,\varphi\gamma)$ as the top row of the commutative diagram

$$(1.8)\qquad \begin{array}{ccccc} e\gamma : N & \overset{\varkappa_o}{\rightarrowtail} & G^\gamma & \overset{\pi_o}{\twoheadrightarrow} & Q_1 \\ \Vert & & \downarrow{\scriptstyle \gamma^\bullet} & & \downarrow{\scriptstyle \gamma} \\ e : N & \overset{\varkappa}{\rightarrowtail} & G & \overset{\pi}{\twoheadrightarrow} & Q \end{array}$$

Here $G^\gamma = \{ (g,q) \in G \times Q_1 \mid \pi g = \gamma q \}$ and $\varkappa_o = \{n \longmapsto (\varkappa n,1)\}$, $\pi_o = \{(g,q) \longmapsto q\}$, $\gamma^\bullet = \{(g,q) \longmapsto g\}$.

This construction is called the <u>backward induced extension</u> $e\gamma$, it makes use of the pull-back in the category of groups (and is valid for arbitrary N, too). The reader easily verifies the assertions implied in the above definition, e.g. that $e\gamma$ is an extension. In terms of factor systems: If $q \longmapsto u_q \in G$ is a transversal for e with associated factor system $f: Q \times Q \to N$, then $q \longmapsto (u_{\gamma q}, q)$ for $q \in Q_1$ is a transversal for $e\gamma$ with factor system $f(\gamma \times \gamma): Q_1 \times Q_1 \to N$. The group G^γ may be known to the reader as Wielandt's product $G \curlywedge Q_1$, the latter being defined when γ is epimorphic.

1.4 PROPOSITION. In the situation of Definition 1.3, any morphism $(\alpha,\beta,\gamma): e_1 \to e$ from an extension e_1 can be factored uniquely as

$$e_1 \xrightarrow{(\alpha,\eta,1)} e\gamma \xrightarrow{(1,\gamma^\bullet,\gamma)} e .$$

Here $1: Q_1 \to Q_1$ and $1: N \to N$ denote the identity homomorphisms. If $e \equiv e'$, then $e\gamma \equiv e'\gamma$. Hence $\gamma^* = \{[e] \longmapsto [e\gamma]\}$: $\mathrm{Opext}(Q,N,\varphi) \longrightarrow \mathrm{Opext}(Q_1,N,\varphi\gamma)$ is well-defined.

PROOF. The map η required for the factorization of $e_1 \to e$ is, and has to be, $\{g \longmapsto (\beta g, \pi_1 g)\}: G_1 \to G^\gamma$. If $(1,\beta,1): e \to e'$ affords the congruence $e \equiv e'$, then the factorization of $(1,\beta\circ\gamma^\bullet,\gamma): e\gamma \to e'$ yields $e\gamma \equiv e'\gamma$. $\square$

1.5 DEFINITION. Given a Q-extension e of the Q-module (N,φ) and a Q-linear map $\alpha\colon N \to (N_2,\varphi_2)$. We define a Q-extension αe of (N_2,φ_2) as the bottom row of the commutative diagram

$$
(1.9)\qquad
\begin{array}{rccccc}
e: & N & \overset{\varkappa}{\rightarrowtail} & G & \overset{\pi}{\twoheadrightarrow} & Q \\
 & {\scriptstyle\alpha}\downarrow & & \downarrow{\scriptstyle\alpha_\bullet} & & \| \\
\alpha e: & N_2 & \overset{\varkappa_0}{\rightarrowtail} & G_\alpha & \overset{\pi_0}{\twoheadrightarrow} & Q
\end{array}
\quad .
$$

Here $G_\alpha = ((N_2,\varphi_2\pi) \rtimes G)/S$ with $S = \{ (\alpha n^{-1},\varkappa n) \mid n \in N \}$ and $\varkappa_0 = \{n \longmapsto (n,1)S\}$, $\pi_0 = \{(n_2,g)S \longmapsto \pi g\}$, $\alpha_\bullet = \{g \longmapsto (1,g)S\}$.

In terms of factor systems: If $q \mapsto u_q$ is a transversal for e and f the associated factor system as above, then $q \longmapsto (1,u_q)S$ for $q \in Q$ is a transversal for αe with factor system $\alpha f\colon Q \times Q \to N_2$.

We call αe a <u>forward induced extension</u>. The reader should verify the assertions implied in the above definition. Indeed, S is a normal subgroup of $N_2 \rtimes G$ since even $\{n \longmapsto (\alpha n^{-1},\varkappa n)\}\colon N \to N_2 \rtimes G$ is a normal monomorphism. The left square of (1.9) is commutative since $(\alpha n,1)$ and $(1,\varkappa n)$ in $N_2 \rtimes G$ are congruent modulo S, for all $n \in N$. The map π_0 is well-defined, αe is indeed an extension, and the associated operation of Q on N_2 is given by φ_2. Thus the extension αe is central precisely when Q acts trivially on N_2; in this case $G_\alpha = (N_2 \times G)/S$.

1.6 PROPOSITION. In the situation of Definition 1.5, any morphism $(\alpha,\beta,\gamma)\colon e \to e_2$ into a Q_2-extension of (N_2,φ_2) can be factored uniquely as

$$e \xrightarrow{(\alpha,\alpha_\bullet,1)} \alpha e \xrightarrow{(1,\xi,\gamma)} e_2 .$$

If $e' \equiv e$, then $\alpha e' \equiv \alpha e$. Hence $\alpha_* = \{[e] \longmapsto [\alpha e]\}$: $\mathrm{Opext}(Q,N,\varphi) \longrightarrow \mathrm{Opext}(Q,N_2,\varphi_2)$ is well-defined.

PROOF. The map ξ required for the factorization is, and has to be, $\{(n_2,g)S \longmapsto \varkappa_2 n_2 \cdot \beta g\}$. This is well-defined since $\{(n_2,g) \longmapsto \varkappa_2 n_2 \cdot \beta g\}: N_2 \rtimes G \longrightarrow G_2$ is a homomorphism which annihilates S. If $(1,\beta,1): e' \to e$ affords the congruence $e' \equiv e$, then the factorization of $(\alpha,\alpha_{*}\beta,1): e' \to \alpha e$ yields $\alpha e' \equiv \alpha e$. □

1.7 REMARK. The morphism $(1,\gamma^{*},\gamma): e\gamma \to e$ used in the definition of $e\gamma$ has $\alpha = 1$. Proposition 1.4 implies that any morphism $(1,\beta,\gamma): e_1 \to e$ exhibits e_1 as (congruent to the induced extension) $e\gamma$. The morphism $(\alpha,\alpha_{*},1): e \to \alpha e$ used in the definition of αe has $\gamma = 1$. Conversely let a morphism $(\alpha,\beta,1): e \to e_2$ of extensions with abelian kernel be given. Then α is Q-homomorphic from (1.5), and Proposition 1.6 implies that e_2 is (congruent to the induced extension) αe. Obviously $e1 \equiv e \equiv 1e$.

1.8 EXAMPLES. The preceding remark often allows one to describe an induced extension without invoking the definition. In the following, let e as in (1.1) be a Q-extension of the Q-module (N,φ). If P is a subgroup of Q and $i: P \hookrightarrow Q$ denotes the inclusion, look at

$$\begin{array}{cccccc} e|P: & N & \xrightarrow{\varkappa'} & \pi^{-1}(P) & \xrightarrow{\pi'} & P \\ & \| & & \downarrow & & \downarrow i \\ & N & \xrightarrow{\varkappa} & G & \xrightarrow{\pi} & Q \end{array} ,$$

where $\varkappa'$ and π' are the (unique) maps which render this diagram commutative. Then the top row $e|P$ is a P-extension of $(N,\varphi \circ i)$ and $[e|P] = i^{*}[e]$. Sometimes i^{*} is called res for "restriction". Next, let (N_2,φ_2) be another Q-module and $i: N \rightarrowtail N \times N_2$ be the natural inclusion as the first summand. Then the following commutative diagram exhibits $i_{*}[e] = [e_2]$,

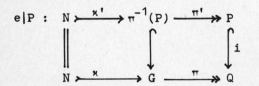

$$\begin{array}{ccccccc}
e\;: & N & \xrightarrow{\;\varkappa\;} & G & \xrightarrow{\;\pi\;} & Q \\
 & \downarrow{\scriptstyle i} & & \downarrow{\scriptstyle \lambda} & & \| \\
e_2\;: & N \times N_2 & \xrightarrow{\;\varkappa_2\;} & (N_2,\varphi_2\pi) \rtimes G & \xrightarrow{\;\pi_2\;} & Q
\end{array}$$

where $\varkappa_2 = \{(n,n_2) \longmapsto (n_2,\varkappa n)\}$, $\pi_2 = \{(n_2,g) \longmapsto \pi g\}$, and $\lambda = \{g \longmapsto (1,g)\}$.

The use of induced extensions via induced factor systems is "folklore" and often implicit in computations. However, if the extension group G is constructed from a factor set as in 1.2, it usually takes a tedious effort to exhibit the properties of G. On the other hand, the diagram constructions for induced extensions are much more helpful in this regard. The model for this section was the treatment that MAC LANE [2; III §1] had given for modules. Actually Mac Lane had used the diagram constructions for groups in a preliminary version of his book.

1.9 DEFINITION. A <u>derivation</u> of the group Q in the Q-module (A,φ) is a set map $d\colon Q \to A$ with $d(x\cdot y) = d(x)\cdot {}^{x}d(y)$ for all x and y in Q. The set of all derivations of Q in A forms an additive abelian group $\mathrm{Der}(Q,A,\varphi)$ with addition + defined by pointwise multiplication. If $\pi\colon G \to Q$ is a group homomorphism, there is an induced homomorphism $\mathrm{Der}(\pi) = \{d \longmapsto d\circ\pi\}$: $\mathrm{Der}(Q,A,\varphi) \longrightarrow \mathrm{Der}(G,A,\varphi\pi)$.

Alternatively, the set map $d\colon Q \to A$ is a derivation of Q in (A,φ) exactly, if $\{q \longmapsto (d(q),q)\}\colon Q \to (A,\varphi) \rtimes Q$ is a group homomorphism. In the case of trivial operation, obviously $\mathrm{Der}(Q,A,0) = \mathrm{Hom}(Q,A)$.

Given Q_i-extensions e_i of (N_i,φ_i) for $i=1,2$ and group homomorphisms $\sigma\colon N_1 \to N_2$ and $\gamma\colon Q_1 \to Q_2$. A basic technical problem is: When can a homomorphism $\beta\colon G_1 \to G_2$ be found as to make the following diagram commutative?

$$\begin{array}{ccccccc} e_1 : & N_1 & \rightarrowtail & G_1 & \twoheadrightarrow & Q_1 \\ (1.10) & \downarrow \alpha & & \downarrow \beta & & \downarrow \gamma \\ e_2 : & N_2 & \rightarrowtail & G_2 & \twoheadrightarrow & Q_2 \end{array}$$

We then say: The diagram can be solved for β. As pointed out above, a necessary condition is that α and γ satisfy (1.5).

1.10 THEOREM. Given the data of diagram (1.10) with α and γ satisfying (1.5), thus αe_1 being defined for the Q_1-homomorphism $\alpha\colon (N_1,\varphi_1) \to (N_2,\varphi_2\gamma)$. Then diagram (1.10) can be solved for β if, and only if, $\alpha e_1 \equiv e_2\gamma$. In this case, the solution set is in bijective correspondence with $\mathrm{Der}(Q_1,N_2,\varphi_2\gamma)$.

This theorem and the following three propositions constitute the <u>calculus of induced extensions</u> and will be frequently used hereafter.

PROOF. If $\alpha e_1 \to e_2\gamma$ is a congruence, then the composition $e_1 \to \alpha e_1 \to e_2\gamma \to e_2$ is a map of type $(\alpha,\cdot,\gamma)$. Conversely, let a morphism $(\alpha,\beta,\gamma)\colon e_1 \to e_2$ be given. We factor (α,β,γ) using Propositions 1.4 and 1.6 in either order and thus obtain an equivalence $(1,\xi,1)\colon \alpha e_1 \to e_2\gamma$. The uniqueness parts of 1.4 and 1.6 imply that the map ξ is uniquely determined by β. Hence the solution set bijectively corresponds to the set of congruences $(1,\cdot,1)\colon$ $\alpha e_1 \to e_2\gamma$ or, likewise, to the set of the self-congruences of e.g. αe_1 . Let $e\colon A \overset{\varkappa}{\rightarrowtail} G \overset{\pi}{\twoheadrightarrow} Q$ be a Q-extension of (A,φ) . If $d\colon Q \to A$ is a derivation, then $\xi_d = \{g \longmapsto \varkappa d(\pi g)\cdot g\}\colon G \to G$ is a homomorphism such that $(1,\xi_d,1)\colon e \to e$ is a morphism of extensions. Conversely, given a self-congruence $(1,\xi,1)\colon e \to e$. The reader checks that the rule $\varkappa d'(g) = \xi(g)\cdot g^{-1}$ for $g \in G$ defines a derivation d' of G in $(A,\varphi\pi)$. Since d' is constant on cosets

$g \cdot \varkappa A$, d' determines a derivation d_ξ of Q in (A,φ) with $d' = d_\xi \circ \pi$. Clearly the assignments $d \longmapsto \xi_d$ and $\xi \longmapsto d_\xi$ are inverse to each other. □

1.11 PROPOSITION. Let (N,φ) be a Q-module and e a Q-extension of N.

(a) If $\alpha_1\colon N \to (N_1,\varphi_1)$ and $\alpha_2\colon N_1 \to (N_2,\varphi_2)$ are Q-homomorphisms, then $\alpha_2(\alpha_1 e) \equiv (\alpha_2\alpha_1)e$.

(b) If $\gamma_1\colon Q_1 \to Q$ and $\gamma_2\colon Q_2 \to Q_1$ are homomorphisms, then $(e\gamma_1)\gamma_2 \equiv e(\gamma_1\gamma_2)$.

(c) If $\alpha\colon N \to (N_1,\varphi_1)$ is a Q-homomorphism and $\gamma\colon Q_1 \to Q$ a homomorphism, then $\alpha(e\gamma)$ is defined with N_1 being treated as the Q_1-module $(N_1,\varphi_1\gamma)$ and $\alpha(e\gamma) \equiv (\alpha e)\gamma$.

PROOF. Apply the "only if"-part of Theorem 1.10 to $e \to \alpha_1 e \to \alpha_2(\alpha_1 e)$ in case (a), to $(e\gamma_1)\gamma_2 \to e\gamma_1 \to e$ in case (b), and to $e\gamma \to e \to \alpha e$ in case (c). □

1.12 PROPOSITION. Given e and γ as in Definition 1.3. The extension $e\gamma$ is split if, and only if, there exists a lifting $\delta\colon Q_1 \to G$ with $\pi\delta = \gamma$.

PROOF. If $\sigma\colon Q_1 \to G^\gamma$ is a splitting of $e\gamma$, then $\beta^\gamma \cdot \sigma$ satisfies $\pi(\beta^\gamma \circ \sigma) = \gamma$. Conversely, let a lifting δ with $\pi\delta = \gamma$ be given. Then $\sigma = \{q \longmapsto (\delta q, q)\}\colon Q_1 \to G^\gamma$ is a splitting of $e\gamma$. □

1.13 PROPOSITION. Given e and α as in Definition 1.5. The extension αe is split if, and only if, there exists a derivation $d\colon G \to (N_2,\varphi_2\pi)$ with $d \circ \varkappa = \alpha$.

Included is the well-known fact that e itself is split precisely when a derivation $d\colon G \to N$ with $d \circ \varkappa = 1_N$ exists (splitting derivation). In case of central extensions, the proposition calls for a homomorphism $d\colon G \to N_2$ extending α. On the other extreme, our proof easily adapts to extensions with non-abelian kernel. In this case, the map $\bar{d} = \{g \longmapsto d(g)^{-1}\}$ rather than d is required to satisfy $\bar{d}(x\cdot y) = \bar{d}(x)\cdot {}^{x}\bar{d}(y)$.

PROOF. If $d\colon G \to N_2$ and $\Sigma\colon G \to G_\alpha$ are set maps satisfying

(1.11) $\qquad \alpha_*(g) = \varkappa_0 d(g)\cdot\Sigma(g) \qquad$ for all $g \in G$,

then (i) $\pi_0\Sigma = \pi$, (ii) $d\varkappa = \alpha$ if and only if $\Sigma\varkappa = 0$, and (iii) d is a derivation if and only if Σ is a homomorphism. The proofs of (i) and (ii) are immediate. The action of $g \in G$ on $(N_2, \varphi_2\pi)$ can also be given by conjugation with $\alpha_*(g)$ in G_α . Using this, we obtain

$$\varkappa_0 d(g\cdot g_1)\cdot\Sigma(g\cdot g_1) = \varkappa_0[{}^{g}d(g_1)\cdot d(g)]\cdot\Sigma g\cdot\Sigma g_1 \quad \text{for all } g, g_1 \in G .$$

Since N_2 is abelian, (iii) follows.

Now to our assertion. Given a splitting $\sigma\colon Q \to G_\alpha$ of αe , set $\Sigma = \sigma\pi$ and define d by (1.11). This is possible since $(\alpha_* g)\cdot(\Sigma g)^{-1} \in \operatorname{Ker} \pi_0 = \operatorname{Im} \varkappa_0$. It follows from (ii) and (iii) that d is a derivation with $d\varkappa = \alpha$. Conversely, let d be a derivation with $d\varkappa = \alpha$. Define Σ by (1.11). Then (ii) and (iii) give that Σ is a homomorphism with $\Sigma\varkappa = 0$, hence Σ factors uniquely as $\Sigma = \sigma\pi$. The map $\sigma\colon Q \to G_\alpha$ splits αe , since $\pi_0\sigma\pi = 1\circ\pi$. □

1.14 PROPOSITION. Let Q be a finite group of order q and (N,φ) a Q-module, let $\bar{q} = \{x \longmapsto x^q\}\colon N \to N$. Then $\bar{q}$ is a Q-module homomorphism and $\bar{q}_*\colon \operatorname{Opext}(Q,N,\varphi) \to \operatorname{Opext}(Q,N,\varphi)$ is the constant map onto the class of the split extensions.

PROOF. Let e as in (1.1) be any Q-extension of (N,φ) . In view of 1.13, it suffices to find a derivation $d\colon G \to (N, \varphi\pi)$ with $d\varkappa = \bar{q}$. Choose any transversal $\{u_q\}$ and define d by

$$\varkappa d(g) = \prod_{r\in Q} C(g,r) \quad , \quad C(g,r) := g\cdot u_r \cdot u^{-1}_{\pi(g)\cdot r} \quad .$$

This is well-defined, since $D(g,r) \in \operatorname{Ker} \pi = \operatorname{Im} \varkappa$. One easily verifies

(1.12) $C(g\cdot h,r) = gC(h,r)g^{-1}\cdot C(g,\pi(h)r)$ for $g,h \in G$ and $r \in Q$.

Fix g and h for the moment and take the product of (1.12) for all $r \in Q$. Since N is an abelian group and the value of a product does not depend on the multiplication index, the formula $d(g\cdot h) = {}^{\pi(g)}d(h)\cdot d(g)$ follows. Hence d is a derivation. Since $C(\varkappa(n),r) = \varkappa(n)$ for $n \in N$, clearly $d\varkappa(n) = n^q$. □

1.15 DEFINITION. Let $\gamma: G_2 \to G_1$ and $\alpha: A_1 \to A_2$ be group homomorphisms and (A_i,φ_i) be G_i-modules for $i=1,2$. We call $(\gamma,\alpha): (G_1,A_1) \to (G_2,A_2)$ a <u>copair</u> if (1.13) is satisfied,

(1.13) $\alpha({}^{\gamma(g)}a) = {}^{g}\alpha(a)$ for all $g \in G_2$, $a \in A_1$.

Actually, (γ,α) is a copair exactly, if $\alpha:(A_1,\varphi_1\gamma) \to (A_2,\varphi_2)$ is G_2-homomorphic. For a copair (γ,α) define $\operatorname{Opext}(\gamma,\alpha) = \alpha_*\gamma^*$: $\operatorname{Opext}(G_1,A_1,\varphi_1) \to \operatorname{Opext}(G_2,A_1,\varphi_1\gamma) \to \operatorname{Opext}(G_2,A_2,\varphi_2)$. The copairs form a category with composition $(\gamma_2,\alpha_2)\circ(\gamma_1,\alpha_1) = (\gamma_1\circ\gamma_2,\alpha_2\circ\alpha_1)$, which is defined whenever $\gamma_1\circ\gamma_2$ and $\alpha_2\circ\alpha_1$ are defined.

1.16 EXERCISE. Show that Opext is a covariant functor from the category of copairs to the category of sets. (Theorem 2.4 below implies that Opext can be made a functor to the category of abelian groups).

1.17 EXAMPLE. Given a group Q and a Q-module (A,φ) . Fix an $h \in Q$ and let $i=\{q \longmapsto h^{-1}qh\}: Q \to Q$ be the inner automorphism and $l=\{a \longmapsto ha\}: A \to A$. The reader verifies that (i,l) is a copair. For every $e \in \operatorname{Opext}(Q,A,\varphi)$ we have a commutative diagram

$$\begin{array}{lccccc} e: & A & \rightarrowtail & G & \twoheadrightarrow & Q \\ & \downarrow{\scriptstyle l^{-1}} & & \downarrow{\scriptstyle j} & & \downarrow{\scriptstyle i} \\ e: & A & \rightarrowtail & G & \twoheadrightarrow & Q \end{array} ,$$

where $j = \{g \longmapsto k^{-1}gk\} : G \to G$ for some fixed $k \in G$ with $\pi k = h$. By Theorem 1.10 this implies $l_*^{-1}[e] = i^*[e]$ in $\mathrm{Opext}(Q,A,\varphi i)$. Thus $\mathrm{Opext}(i,l)$ is the identity map.

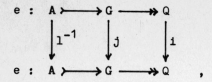

2. The Exact Sequence for Opext

The main purpose of this section is to derive the 5-term exact sequence (2.3), associated with a group extension $N \rightarrowtail G \twoheadrightarrow Q$ and a Q-module A. This sequence has found numerous important applications. At first, it was obtainable only as the specialization of an exact sequence for cohomology groups that HOCHSCHILD/SERRE [1] had derived with spectral sequence techniques. However, in the special case that G is a free group, this sequence was known to EILENBERG/MAC LANE [2; Thm. 13.1] and used to describe $H^2(Q,A)$. Later sequence (2.3) was constructed just from the axioms for derived functors, a certain final stage being reached by HILTON/STAMMBACH [1; VI §10].

We feel that the most natural approach to this sequence (2.3) is by group extensions. Then the terms are Opext rather than cohomology groups, all maps are very explicit, and the proof of exactness is a straightforward application of Section 1. These notes will show how this sequence can be used to obtain group theoretical results. In Section 5 we shall compare our presentation with the cohomology version mentioned above. This transition enables the reader to connect sequence (2.3) with applications published elsewhere in terms of group cohomology. In particular, it allows for additional means of computing the Opext groups.

If A and B are subgroups of a group G, let $[A,B]$ denote the subgroup generated by all commutators $[a,b] = aba^{-1}b^{-1}$ with $a \in A$ and $b \in B$. If A and B are normal in G, then so is $[A,B]$. In particular, $[G,G] \trianglelefteq G$ and $G_{ab} := G/[G,G]$ is the <u>commutator quotient</u> of G.

2.1 DEFINITION. For an extension e define the <u>abelianization</u> $ab(e)$ and the <u>centralization</u> $c(e)$ by the following diagram

$$
(2.1)\qquad
\begin{array}{rcccccc}
e: & N & \overset{\varkappa}{\rightarrowtail} & G & \overset{\pi}{\twoheadrightarrow} & Q \\
 & \downarrow{\scriptstyle ab} & & \downarrow{\scriptstyle nat} & & \Vert \\
ab(e): & N_{ab} & \overset{\varkappa'}{\rightarrowtail} & G/\varkappa[N,N] & \overset{\pi}{\twoheadrightarrow} & Q \\
 & \downarrow{\scriptstyle c} & & \downarrow{\scriptstyle nat} & & \Vert \\
c(e): & N/\varkappa^{-1}[G,\varkappa N] & \overset{\varkappa''}{\rightarrowtail} & G/[G,\varkappa N] & \overset{\pi}{\twoheadrightarrow} & Q
\end{array}
$$

where the vertical maps are natural maps onto factor groups and the horizontal maps are induced by $\varkappa$ and π. (We usually write $[G,N]$ in place of $\varkappa^{-1}[G,\varkappa N]$ or $[G,\varkappa N]$, if the context permits.)

The extension $ab(e)$ with abelian kernel exhibits N_{ab} as a Q-module, the action is induced from conjugation in G. The extension $c(e)$ is central. Clearly any morphism $e \to e_1$ of extensions factors uniquely over $e \to ab(e)$ as in (2.1) if e_1 has an abelian kernel; the map $e \to e_1$ factors uniquely over $e \to c(e)$ if e_1 is even central. Hence any morphism $(\alpha,\beta,\gamma)\colon e_1 \to e_2$ as in (1.4) induces maps $(\alpha_{ab},\cdot,\gamma)\colon ab(e_1) \to ab(e_2)$ and $(\alpha_c,\cdot,\gamma)$: $c(e_1) \to c(e_2)$. Proposition 1.6 implies $c(e) \equiv c(ab(e))$. If $e \equiv e'$, clearly $ab(e) \equiv ab(e')$ and $c(e) \equiv c(e')$.

2.2 LEMMA. Let $e_o\colon R \hookrightarrow F \twoheadrightarrow Q$ be a <u>free presentation</u> of Q, i.e. F is a free group and $R = Ker(F \to Q)$. Then $\bar{e}_o = ab(e_o)$ has the following property: for every Q-extension e of any (N,φ) there is a Q-homomorphism $f : R_{ab} \to N$ such that $e \equiv f(\bar{e}_o)$ or, equivalently, $[e] = f_*[ab(e_o)]$.

PROOF. Let $e\colon N \rightarrowtail G \twoheadrightarrow Q$ be any extension with N abelian. Then $F \to Q$ can be lifted over $G \twoheadrightarrow Q$ by the freeness of F; we obtain a morphism $(f',\cdot,1)\colon e_o \to e$ of extensions. As N is abelian, this morphism can be factored over $e_o \to ab(e_o)=\bar{e}_o$ and yields a commutative diagram

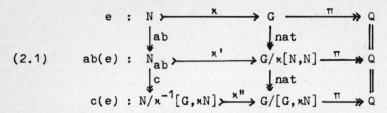

$$\begin{array}{ccccccc}
\bar{e}_o : & R_{ab} & \rightarrowtail & F/[R,R] & \twoheadrightarrow & Q \\
 & \downarrow{\scriptstyle f} & & \downarrow & & \Vert \\
e\ : & N & \rightarrowtail & G & \twoheadrightarrow & Q
\end{array}$$

Then f is Q-homomorphic by (1.5) and $e \equiv f(\bar{e}_o)$ by 1.7. Note that $\bar{e}_o$ depends only on the choice of e_o, but not on e. □

2.3 NOTATION. If Q is a group, then the <u>diagonal</u> $\Delta = \{q \longmapsto (q,q)\}: Q \to Q \times Q$ is a group homomorphism. If N is an abelian group resp. a Q-module, then the <u>codiagonal</u> $\nabla = \{(a,b) \longmapsto a \cdot b\}: N \times N \to N$ is a homomorphism resp. a Q-linear map.

Let Q be abelian for the moment. Then our definitions of induced extensions αe and $e\gamma$ in 1.5 and 1.3 and of induced maps α_* and γ^* in 1.6 and 1.4 agree on the subset of abelian extensions with the corresponding notions customarily given for Ext(Q,N). In Ext(Q,N) addition is defined by

$$[e_1] + [e_2] = [\nabla_N(e_1 \oplus e_2)\Delta_Q]$$

and makes Ext an additive abelian group. From now on, Q may be noncommutative again. Given any extensions $e_i: N_i \rightarrowtail G_i \twoheadrightarrow Q_i$ for $i=1,2$, let $e_1 \times e_2$ denote the obvious extension $N_1 \times N_2 \rightarrowtail G_1 \times G_2 \twoheadrightarrow Q_1 \times Q_2$. Then

(2.2) $\qquad e_1 + e_2 = \nabla_N((e_1 \times e_2)\Delta_Q)$

makes sense for Q-extensions e_i of one and the same Q-module (N,φ) and is again such an extension: it is called the <u>Baer sum</u>.

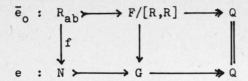

2.4 THEOREM. Let Q be a group and (N,φ) a Q-module. Then $\mathrm{Opext}(Q,N,\varphi)$ becomes an additive abelian group if addition is defined by $[e_1] + [e_2] = [e_1 + e_2]$. The class $[e_o]$ of the split extension is the neutral element, and $0_*[e] = [e_o]$ for the zero-map $0\colon N \to N$ and all e. The inverse of $[e]$ is $[(-1_N)e]$. For Q-linear maps $\alpha,\alpha_1,\alpha_2 : N \to (N_1,\varphi_1)$ and group homomorphisms $\gamma\colon Q_1 \to Q$ the maps $\alpha_* = \{[e] \longmapsto [\alpha e]\}\colon \mathrm{Opext}(Q,N,\varphi) \to \mathrm{Opext}(Q,N_1,\varphi_1)$ and $\gamma^* = \{[e] \longmapsto [e\gamma]\}\colon \mathrm{Opext}(Q,N,\varphi) \to \mathrm{Opext}(Q_1,N,\varphi\gamma)$ are homomorphisms, and moreover $(\alpha_1 + \alpha_2)_* = \alpha_{1*} + \alpha_{2*}$.

On the inverse. If e is a Q-extension of (N,φ) as in (1.1), let $\bar{e}\colon N \overset{-\varkappa}{\rightarrowtail} G \overset{\pi}{\twoheadrightarrow} Q$ where $-\varkappa$ denotes the homomorphism $\{n \longmapsto \varkappa(n)^{-1}\}$. Then $[\bar{e}] = -[e]$, because $(-1_N, 1_G, 1_Q)\colon e \to \bar{e}$ is a morphism of extensions.

On the sum. Let e_i be Q-extensions of (N,φ) for $i=1,2$, to which the factor systems $f_i\colon Q \times Q \to N$ correspond. Then $e_1 \times e_2$ obviously allows the factor system $f_1 \times f_2$. By our remarks after Definition 1.3 and Definition 1.5, $e_1 + e_2$ has $\nabla_N(f_1 \times f_2)\Delta_{Q\times Q} = f_1 + f_2$ as one of its factor systems. Hence the addition of extensions is induced by the pointwise addition of factor systems.

PROOF. In the special case of central extensions, the reasoning of MAC LANE [2; Thm. III.2.1, 2nd proof p.70] applies literally. We adapt it to the present situation. Congruences $e_1 \equiv \bar{e}_1$ and $e_2 \equiv \bar{e}_2$ clearly imply a congruence $e_1 \times e_2 \equiv \bar{e}_1 \times \bar{e}_2$, whence $e_1 + e_2 \equiv \bar{e}_1 + \bar{e}_2$. Thus $[e_1] + [e_2]$ is well-defined.

(a) Claim: $\alpha_1 e + \alpha_2 e \equiv (\alpha_1+\alpha_2)e$ for $\alpha_1,\alpha_2\colon N \to N_1$. To this end, the direct product of $e \to \alpha_1 e$ and $e \to \alpha_2 e$ gives a morphism $(\alpha_1 \times \alpha_2,\cdot,1)\colon e \times e \to (\alpha_1 e) \times (\alpha_2 e)$ of extensions. Moreover $(\Delta,\Delta,\Delta)\colon e \to e \times e$ is a morphism of extensions. By 1.7, we conclude that $(\alpha_1 \times \alpha_2)(e \times e)$ is defined and $(\alpha_1 \times \alpha_2)(e \times e) \equiv (\alpha_1 e) \times (\alpha_2 e)$ and $\Delta e \equiv (e \times e)\Delta$. Next Proposition 1.11 gives $\alpha_1 e + \alpha_2 e = \nabla((\alpha_1 e \times \alpha_2 e)\Delta) \equiv \nabla(\{(\alpha_1 \times \alpha_2)(e \times e)\}\Delta) \equiv$

$\equiv (\nabla(\alpha_1 \times \alpha_2))((e \times e)\Delta) \equiv (\nabla(\alpha_1 \times \alpha_2))(\Delta e) \equiv (\nabla(\alpha_1 \times \alpha_2)\Delta)e =$
$= (\alpha_1 + \alpha_2)e$.

(b) Claim: $0_* e \equiv e_o$. With $\beta(g) = (1, \pi(g))$, the following diagram is commutative and gives the assertion by 1.6:

$$\begin{array}{rcccccc}
e : & N & \rightarrowtail & G & \xrightarrow{\pi} & Q & \\
 & \downarrow 0 & & \downarrow \beta & & \| & \\
e_o : & N & \xrightarrow{\kappa_o} & N \rtimes Q & \xrightarrow{\pi_o} & Q & .
\end{array}$$

(c) Commutative and associative laws. Let Q-extensions e_1 and e_2 of (N, φ) be given. By 2.2, there are $\bar{e}_o$ and homomorphisms $f_i \colon R_{ab} \to N$ such that $e_i \equiv f_i(\bar{e}_o)$. Hence by step (a), $e_1 + e_2 \equiv$
$\equiv f_1(\bar{e}_o) + f_2(\bar{e}_o) \equiv (f_1 + f_2)(\bar{e}_o) = (f_2 + f_1)(\bar{e}_o) \equiv \ldots \equiv e_2 + e_1$.
The associative law can be proved in the same spirit.

(d) Claim: $\alpha(e_1 + e_2) \equiv \alpha e_1 + \alpha e_2$ and $(e_1 + e_2)\gamma \equiv e_1\gamma + e_2\gamma$. Again, let $e_i \equiv f_i(\bar{e}_o)$ by 2.2. Then by 1.11 and step a) above,
$\alpha e_1 + \alpha e_2 \equiv \alpha f_1(\bar{e}_o) + \alpha f_2(\bar{e}_o) \equiv (\alpha f_1 + \alpha f_2)\bar{e}_o = (\alpha(f_1 + f_2))\bar{e}_o \equiv$
$\equiv \alpha((f_1 + f_2)\bar{e}_o) \equiv \alpha(e_1 + e_2)$ and $e_1\gamma + e_2\gamma = (f_1\bar{e}_o)\gamma + (f_2\bar{e}_o)\gamma \equiv$
$f_1(\bar{e}_o\gamma) + f_2(\bar{e}_o\gamma) \equiv (f_1 + f_2)(\bar{e}_o\gamma) \equiv ((f_1 + f_2)\bar{e}_o)\gamma \equiv (e_1 + e_2)\gamma$.

Steps (c) and (d) can also be proved by a suitable diagram play, without invoking Lemma 2.2. When adapting the arguments of MAC LANE [2; Thm. III.2.1, 2nd proof] from Ext to Opext , one must proceed with caution for the following reason. The congruence $\alpha(e\gamma) \equiv (\alpha e)\gamma$ holds whenever αe is defined, i.e. the quotient group $\bar{Q}$ in e operates on the range of α such that α is $\bar{Q}$-linear. But it may well be that $\alpha(e\gamma)$ is defined and αe is not; e.g. $\nabla(e \times e)$ in the situation of formula (2.2) cannot be defined unless Q operates trivially on N. $\square$

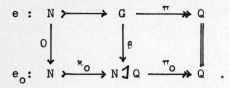

2.5 COROLLARY. Given a direct sum diagram

$$N_1 \underset{p_1}{\overset{i_1}{\rightleftarrows}} N \underset{i_2}{\overset{p_2}{\rightleftarrows}} N_2$$

of Q-modules, i.e. $p_1 i_1 = 1$ on N_1, $p_2 i_2 = 1$ on N_2, $p_1 i_2 = 0 = p_2 i_1$, and $i_1 p_1 + i_2 p_2 = 1$ on N. (This implies $N \cong N_1 \times N_2$). Then

$$\mathrm{Opext}(Q,N_1) \underset{p_{1*}}{\overset{i_{1*}}{\rightleftarrows}} \mathrm{Opext}(Q,N) \underset{i_{2*}}{\overset{p_{2*}}{\rightleftarrows}} \mathrm{Opext}(Q,N_2)$$

is again a direct sum diagram, where $p_{1*} = \mathrm{Opext}(1_Q, p_1)$ etc.

PROOF. From Definition 1.5, Theorem 2.4, and Proposition 1.11 we have $1_* = 1$, $0_* = 0$, $(\alpha + \beta)_* = \alpha_* + \beta_*$, and $(\alpha \circ \beta)_* = \alpha_* \circ \beta_*$, whence the assertion. □

2.6 EXAMPLE. Given a finite group Q of order q and a finite Q-module (N,φ) of order n. Let P be the set of primes dividing q, and P' the set of the other primes. Then N as an abelian group has a unique Hall P-subgroup N_1 and a unique Hall P'-subgroup N_2, and $N = N_1 \times N_2$. Since every automorphism of N must send N_1 to N_1 and N_2 to N_2, actually $N = N_1 \times N_2$ as Q-modules. Let $\bar{q}: N_2 \to N_2$ be the q-th power map. Then $\bar{q}$ is an automorphism of Q-modules, hence so is $\bar{q}_*: \mathrm{Opext}(Q,N_2) \to \mathrm{Opext}(Q,N_2)$. Since Proposition 1.14 gives $\bar{q}_* = 0$, we conclude $\mathrm{Opext}(Q,N_2) = 0$. Thus $i_{1*}: \mathrm{Opext}(Q,N_1) \to \mathrm{Opext}(Q,N)$ is an isomorphism, in the notation of 2.5. We now interpret i_{1*} by Example 1.8 and conclude that every Q-extension of (N,φ) has a semi-direct factor isomorphic to N_2.

2.7 THEOREM. Given an extension e as in (1.1) and a Q-module (A,φ). Then

$$(2.3)\quad \begin{array}{l} 0 \longrightarrow Der(Q,A,\varphi) \xrightarrow{Der\ \pi} Der(G,A,\varphi\pi) \xrightarrow{\rho(e,A)} Hom_Q(N_{ab},A) \\ \xrightarrow{\theta^*(e,A)} Opext(Q,A,\varphi) \xrightarrow{\pi^*} Opext(G,A,\varphi\pi) \end{array}$$

is a natural exact sequence of abelian groups and homomorphisms, where $\rho = \rho(e,A)$ is such that $\rho(d)$ for $d \in Der(G,A,\varphi\pi)$ is the Q-linear map $\{n\cdot[N,N] \longmapsto d\varkappa(n)\}$, and $\theta^*(e,A) = \{f \longmapsto f_*[ab(e)]\}$.

What does naturality mean? Given a morphism $(\alpha,\beta,\gamma)\colon e_1 \to e_2$ of extensions as in (1.4) and Q_i-modules (A_i,φ_i) for $i=1,2$ and a Q_1-linear map $h\colon (A_2,\varphi_2\gamma) \to (A_1,\varphi_1)$. Then we have a sequence (2.3) from the data $\{e_1,A_1\}$, say s_1, and a similar sequence s_2 from the data $\{e_2,A_2\}$ and maps from each term of s_2 to the corresponding term of s_1; altogether a laddershaped diagram $s_2 \to s_1$. The assertion "(2.3) is natural" means that this ladder always is a commutative diagram. The interesting part of the assertion is that diagram (2.4) be commutative:

$$(2.4)\quad \begin{array}{ccccc} Hom_{Q_1}(N_{1ab},A_1) & \xrightarrow{\theta^*(e_1,A_1)} & Opext(Q_1,A_1,\varphi_1) & \xrightarrow{\pi_1^*} & Opext(G_1,A_1,\varphi_1\pi_1) \\ \uparrow{\scriptstyle Hom(\alpha_{ab},h)} & & \uparrow{\scriptstyle Opext(\gamma,h)} & & \uparrow{\scriptstyle Opext(\beta,h)} \\ Hom_{Q_2}(N_{2ab},A_2) & \xrightarrow{\theta^*(e_2,A_2)} & Opext(Q_2,A_2,\varphi_2) & \xrightarrow{\pi_2^*} & Opext(G_2,A_2,\varphi_2\pi_2) \end{array}$$

PROOF of 2.7. Keep in mind that A is abelian and N operates trivially on A.

(a) Claim: If $nat\colon G \twoheadrightarrow G'=G/\varkappa[N,N]$ is as in 2.1, then $\xi=Der(nat)\colon Der(G',(A,\varphi\pi')) \longrightarrow Der(G,(A,\varphi\pi))$ is an isomorphism. Obviously ξ is a monomorphism. It suffices to show that every derivation $d\colon G \to A$ is constant on the cosets $g\varkappa[N,N]$, hence is in the image of ξ. Since $d\varkappa\colon N \to A$ is a homomorphism, $d\varkappa[N,N] = \{1\}$ and $d(n\cdot g) = d(n)\cdot{}^{n}d(g) = d(g)$ for $g \in G$ and $n \in \varkappa[N,N]$.

(b) For the moment, consider sequence (2.3') which is sequence (2.3) for the extension $ab(e)$ rather than e. Here $\rho' = \rho(ab(e),A)$ is clearly well-defined as a map to $Hom(N_{ab},A)$ and is a homomorphism. It follows from $d(g)\cdot{}^{g}d(g^{-1}) = d(1) = 1$ for $d \in Der(G',A)$ and $g \in G'$, that $\rho'(d)$ is Q-homomorphic. The reader now easily verifies that the sequence (2.3') is exact at $Der(Q,A,\varphi)$ and $Der(G',A,\varphi\pi')$.

(c) We observe that $Der(\pi) = \xi\, Der(\pi')$ and $\rho = \rho'\xi^{-1}$. Hence ρ is well-defined, and (b) implies that sequence (2.3) is exact at $Der(Q,A,\varphi)$ and $Der(G,A,\varphi\pi)$. Theorem 2.4 immediately gives that $\theta^*(e,A)$ and π^* are homomorphisms. We will show below the exactness at $Hom_Q(N_{ab},A)$ and $Opext(Q,A,\varphi)$ and the commutativity of (2.4).

(d) Exactness at $Hom_Q(N_{ab},A)$. The zero element of $Opext(Q,A,\varphi)$ is the class of the split extensions. Hence Proposition 1.13, when applied to the abelian extension $ab(e)$, can be paraphrased as $Im\,\rho' = Ker\,\theta^*(ab(e),A)$. Hence $Im\,\rho = Im(\rho'\xi^{-1}) = Im\,\rho' =$ $= Ker\,\theta^*(ab(e),A) = Ker\,\theta^*(e,A)$.

(e) Exactness at $Opext(Q,A,\varphi)$. If $f \in Hom_Q(N_{ab},A)$, then $\pi^*\theta^*(e,A)f = (f_*ab(e))\pi^*) = f_*(ab(e)\pi^*)$ by Proposition 1.11 and $ab(e)\pi^* = 0$ by Proposition 1.12. Thus $\pi^*\circ\theta^*(e,A) = 0$. Conversely, let $e_1\colon (A,\varphi) \overset{\varkappa_1}{\rightarrowtail} G_1 \overset{\pi_1}{\twoheadrightarrow} Q$ with $\pi^*[e_1] = [e_1\pi] = 0$ be given. Proposition 1.12 yields a map $\delta\colon G \to G_1$ with $\pi_1\delta = \pi$. Let $\alpha\colon N \to A$ be the restriction of δ, then we have a morphism $(\alpha,\delta,1)\colon e \to e_1$ of extensions and hence by 2.1 also $(\alpha_{ab},\cdot,1)\colon ab(e) \to e_1$. Now Theorem 1.10 exhibits $e_1 \equiv \alpha_{ab*}ab(e)$, thus $[e_1] = \theta^*(e,A)\alpha_{ab}$.

(f) Claim: The left square of (2.4) is commutative. From the given morphism $(\alpha,\beta,\gamma)\colon e_1 \to e_2$ we obtain a morphism $(\alpha_{ab},\cdot,\gamma)\colon ab(e_1) \to ab(e_2)$ by 2.1, and Theorem 1.10 yields

$\alpha_{ab}(ab(e_1)) \equiv ab(e_2)\gamma$. For $f \in Hom_{Q_2}(N_{2ab},A_2)$ thus $\theta^*(e_1,A_1)Hom(\alpha_{ab},h)f = (hf)_*[\alpha_{ab}ab(e_1)] = (hf)_*\gamma^*[ab(e_2)] = h_*\gamma^*[f\ ab(e_2)] = Opext(\gamma,h)\theta^*(e_2,A_2)f$.

(g) Claim: The right square of (2.4) is commutative. Indeed, $Opext(\beta,h)\circ\pi_2^* = h_*\beta^*\pi_2^* = h_*(\pi_2\beta)^* = h_*(\gamma\pi_1)^* = h_*\pi_1^*\gamma^* = \pi_1^*h_*\gamma^* = \pi_1^*\circ Opext(\gamma,h)$. □

3. The Schur Multiplicator and the Universal Coefficient Theorem

Now we define the Schur multiplicator $M(Q)$ of a (possibly infinite) group Q and, likewise, the Schur multiplicator $M(\gamma)$ of a homomorphism γ. We define $M(Q)$ by the Schur-Hopf Formula $M(Q) = R\cap[F,F])/[R,F]$, where $R \hookrightarrow F \twoheadrightarrow Q$ is any free presentation of Q. This formula was discovered by SCHUR [2; p.101] for (certain free presentations of) finite groups Q and obtained by HOPF [1] for arbitrary Q in a quite different (topological) context. A first result is that every group extension $(\kappa,\pi)\colon N \rightarrowtail G \twoheadrightarrow Q$ determines a 5-term exact sequence (3.4), in which $M(\pi)\colon M(G) \to M(Q)$ appears. This theorem is quite powerful, although its proof is just applying the isomorphism theorems of elementary group theory. Part of sequence (3.4) was known to SCHUR [1; Satz II,p.31]. But it was only via the Hochschild-Serre spectral sequence that its general form was first discovered and its value appreciated, cf. STALLINGS [1] and STAMMBACH [1].

We then combine this sequence with the exact Der-Hom-Opext sequence (2.3) and prove a group-theoretical version of the Universal Coefficient Theorem (in dimension two), a celebrated theorem in algebraic topology. In the present formulation of this theorem, all maps are given by explicit constructions, while the proof is quite elementary and self-contained.

We wish to add some comments on the use of free presentations in the definition of $M(Q)$. We regard the free presentations e as coordinate systems, each e determining an abelian group $M(Q)_e$. For each pair of free presentations e and $\bar{e}$, there is a unique coordinate isomorphism $M(Q)_e \cong M(Q)_{\bar{e}}$. It will be a great advantage that we may choose a free presentation particularly suited to the given problem. We could have avoided coordinate transformations and any logical difficulties by distinguishing one free presentation for each group Q, e.g. the standard free presentation of Q as in 3.3. Such an approach would be inconvenient e.g. when we wish to discuss the relationship between $M(Q)$ and the deficiency of groups, see Section IV.1. Alternatively, some authors define $M(Q)$ as a certain cohomology group. This is mostly not appropriate, however, for the treatment of infinite groups Q, cf. the remark after II.3.10. The coordinate transformations are easy to handle, anyway, with the techniques introduced in this section.

3.1 LEMMA. There is a functor m from the category of extensions to that of abelian groups, which assigns to the extension e as in (1.1) the abelian group $\varkappa N \cap [G,G]/[\varkappa N,G]$ and to a morphism $(\alpha,\beta,\gamma)\colon e_1 \to e_2$ as in (1.4) the homomorphism induced by β on the subquotients $\varkappa_i N_i \cap [G_i,G_i]/[\varkappa_i N_i,G_i]$ of G_i, for $i=1,2$. Actually, $m(\alpha,\beta,\gamma)$ depends only on γ rather than β.

PROOF. Clearly m is a functor. As for the last assertion, let $(\bar\alpha,\bar\beta,\gamma)\colon e_1 \to e_2$ be another morphism of extensions with the same γ. Then for every $g \in G_1$, there is an $n(g) \in \varkappa_2 N_2$ with $\bar\beta(g) = \beta(g)\cdot n(g)$. Since $\varkappa_2 N_2/[\varkappa_2 N_2,G_2]$ lies in the center of $G_2/[\varkappa_2 N_2,G_2]$, we obtain $[\bar\beta g_1,\bar\beta g_2] \equiv [\beta g_1,\beta g_2]$ modulo $[\varkappa_2 N_2,G_2]$, whence the assertion. □

Given extensions $e_i\colon N_i \rightarrowtail G_i \twoheadrightarrow Q_i$ for $i=1,2$ and a homomorphism $\gamma\colon Q_1 \to Q_2$. If G_1 is a free group, γ certainly can be lifted to a morphism $(\alpha,\beta,\gamma)\colon e_1 \to e_2$ of extensions. By Lemma 3.1 $m(\alpha,\beta,\gamma)$ depends on γ only, not on the choice of β. The following definition of the Schur multiplicator looks artificial, at first, a coordinate-free but more abstract characterization is given in Remark 3.10 (a).

3.2 DEFINITION. If $e: R \rightarrowtail F \twoheadrightarrow Q$ is a free presentation of the group Q, i.e. an extension with a free group F and kernel R, then define the value of the Schur multiplicator M(Q) at e as the abelian group $M(Q)_e = R \cap [F,F]/[R,F]$. Given a homomorphism $\gamma: Q \to Q'$ and free presentations e and e' of Q and Q', respectively, define $M(\gamma)_{e|e'} = m(\alpha,\beta,\gamma)$ for any choice of lifting (α,β,γ) of γ. Then

(3.1) $M(1_Q)_{e|e} = 1$ and $M(\gamma_1\gamma_2)_{e|e''} = M(\gamma_1)_{e'|e''} \circ M(\gamma_2)_{e|e'}$,

whenever $\gamma_1\gamma_2$ is defined. Free presentations of Q certainly exist. We regard the free presentation e as a coordinate system for M(Q) and the isomorphisms $M(1)_{e|e'}$ as coordinate transformations. Thus every homomorphism $\gamma : Q \to Q'$ induces a well-defined homomorphism $M(\gamma): M(Q) \to M(Q')$. The defining formula for M(Q) is called the Schur-Hopf Formula.

PROOF of implied properties. The formulas (3.1) follow by using the obvious liftings. The map $M(1)_{e|e'}$ is an isomorphism, because $M(1)_{e'|e}$ is its inverse. The properties (3.1) also imply that the various maps $M(\gamma)$ are compatible with the coordinate transformations. □

Equations (3.1) mean that M(-) is a functor from (the category of) groups to abelian groups, once you have chosen a fixed free presentation for each group Q. For example, let $G = N \rtimes Q$ be a semidirect product with splitting $\sigma: Q \to G$ (notation as in (1.3)), but N may be a non-abelian group with operators Q rather than a Q-module. Then the composite map

$$M(Q) \xrightarrow{M(\sigma)} M(G) \xrightarrow{M(\pi_o)} M(Q)$$

is identity due to $\pi_o \circ \sigma = 1_Q$, thus $M(G) \cong M(Q) \times \operatorname{Ker} M(\pi_o)$ is a direct product. For details on the decomposition of $M(N \rtimes Q)$ see HAEBICH [2], in view of Proposition 5.5 also EVENS [2] and TAHARA [1].

3.3 DEFINITION. For a group Q, let the <u>standard free presentation</u> of Q be

$$(3.2) \qquad e(Q) : R_Q \hookrightarrow F_Q \xrightarrow{\pi} Q ,$$

where F_Q is the free group on $\{X_q \mid q \in Q, q \neq 1\}$ and π maps X_q onto q and R_Q is the kernel of π. Also write X_1 for 1 in F_Q. A homomorphism $\gamma: Q \to Q'$ induces a morphism $(\cdot, \beta, \gamma): e(Q) \to e(Q')$ with $\beta(X_q) = X'_{\gamma q}$.

3.4 LEMMA. If Q is free or cyclic, then $M(Q) = 0$.

PROOF. There is an abvious free presentation with $R = 0$ if Q is free, and another with $F = \mathbb{Z}$ abelian if Q is cyclic. □

3.5 PROPOSITION (cf. STALLINGS [1; p.172], STAMMBACH [1; p.170]). Every extension e as in (1.1) determines an exact sequence

$$(3.3) \qquad M(G) \xrightarrow{M(\pi)} M(Q) \xrightarrow{\theta_*(e)} \frac{N}{[N,G]} \xrightarrow{\varkappa'} G_{ab} \xrightarrow{\pi_{ab}} Q_{ab} \longrightarrow 0 ,$$

natural with respect to morphisms of extensions. (Here $\varkappa'$ is induced by $\varkappa: N \to G$.) If $e \equiv e'$, then $\theta_*(e) = \theta_*(e')$. If $G = F$ is free and $\varkappa$ an inclusion, then $\theta_*(e)$ is just the inclusion map $M(Q)_e = (N \cap [F,F])/[N,F] \hookrightarrow N/[N,F]$.

Recall that $N/[N,G]$ stands for $N/\varkappa^{-1}[\varkappa N,G]$. The naturality assertion implies $\theta_*(e\gamma) = \theta_*(e)M(\gamma)$ and $\theta_*(\alpha e) = \alpha_c \theta_*(e)$, with α_c being defined in 2.1. The homomorphism $\theta_*(e)$ is sometimes called the "homology transgression".

PROOF. Let $S \lhd F \xrightarrow{\rho} G$ be a free presentation of G and $R = \mathrm{Ker}(\pi\rho) \supseteq S$. Thus we have a commutative diagram

$$(3.4) \qquad \begin{array}{ccccccc} e'' : & S & \hookrightarrow & F & \xrightarrow{\rho} & G \\ & \downarrow & & \| & & \downarrow{\scriptstyle \pi} \\ e' : & R & \hookrightarrow & F & \xrightarrow{\pi\rho} & Q \\ & \downarrow{\scriptstyle \sigma} & & \downarrow{\scriptstyle \rho} & & \| \\ e\ : & N & \xrightarrow{\varkappa} & G & \xrightarrow{\pi} & Q \end{array} ,$$

in which the rows and the left column are short-exact. In the top line of the following diagram, let all maps be induced by inclusions of the pairs (numerator, denominator), while the isomorphism σ' is induced by σ:

$$\begin{array}{ccccc} \dfrac{S \cap [F,F]}{[S,F]} & \longrightarrow & \dfrac{R \cap [F,F]}{[R,F]} & \longrightarrow & \dfrac{R}{S\cdot[R,F]} \\ \| & & \| & & \cong \downarrow \sigma' \\ M(G)_{e''} & \xrightarrow{M(\pi)_{e''|e'}} & M(Q)_{e'} & \xrightarrow{\theta_*(e)} & \dfrac{N}{[N,G]} \end{array}$$

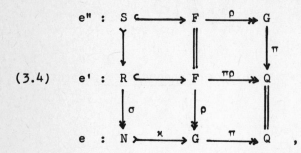

The left square is commutative because of (3.4), we define $\theta_*(e)$ as to make the right square commutative also. The given description of $\theta_*(e)$ in case $G = F$ free is immediate. Since $[F,F] \cap (S\cdot[R,F]) = ([F,F] \cap S)\cdot[R,F]$ by the modular law, $\mathrm{Im}\, M(\pi) = \mathrm{Ker}\, \theta_*(e)$. Likewise $R \cap ([F,F]\cdot S) = (R \cap [F,F])\cdot S$ implies $\mathrm{Im}\, \theta_*(e) = \mathrm{Ker}\, (N/[N,G] \to G_{ab})$. Sequence (3.3) is natural in e, since every morphism of extensions can be extended to a translation of the associated diagrams (3.4), whatever the arbitrary choices were. A similar argument shows that the definition of $\theta_*(e)$ is independent from the free presentation e' chosen as a coordinate system for $M(Q)$. The remaining assertion is immediate from naturality. □

3.6 ADDENDA. If e is central, we confuse N/O , N_{ab} , and N with each other and obtain the exact sequence

$$(3.3') \qquad M(G) \xrightarrow{M(\pi)} M(Q) \xrightarrow{\theta_*(e)} N \xrightarrow{\varkappa_{ab}} G_{ab} \xrightarrow{\pi_{ab}} Q_{ab} \longrightarrow 0 \ ,$$

natural with respect to morphisms of central extensions. The formulation of Proposition 3.5 determines $\theta_*(e)$ completely, although this is not apparent at first. Let $\bar{e}: R \hookrightarrow F \twoheadrightarrow Q$ be the free presentation of Q, at which $M(Q)$ shall be evaluated. Since F is free, we obtain a commutative diagram

$$(3.5) \qquad \begin{array}{cccccc} \bar{e}: & R & \hookrightarrow & F & \twoheadrightarrow & Q \\ & \downarrow{\scriptstyle \alpha} & & \downarrow{\scriptstyle \beta} & & \| \\ e: & N & \xrightarrow{\varkappa} & G & \twoheadrightarrow & Q \end{array}$$

for some suitable maps β and α, possibly not surjective. The naturality of θ_* , when applied to $(\alpha,\beta,1): \bar{e} \to e$, exhibits $\theta_*(e)$ as the composite

$$M(Q)_{\bar{e}} = (R \cap [F,F])/[R,F] \xhookrightarrow{\theta_*(\bar{e})} R/[R,F] \xrightarrow{\alpha_c} N/\varkappa^{-1}[\varkappa N,G] \ .$$

Here α_c is induced by α as in 2.1.

3.7 LEMMA. Let (A,φ) be a Q-module and $e: N \rightarrowtail G \twoheadrightarrow Q$ an extension. If $\psi: N_{ab} \to A$ is a Q-homomorphism, then

$$\theta_*\theta^*(e,A)\psi = \theta_*[\psi\, ab(e)] = \psi_c\theta_*(e) \ : \ M(Q) \to N/[N,G] \to A_Q \ .$$

Here $A_Q = A \,/\, \{\, {}^{q}a \cdot a^{-1} \mid a \in A \ , \ a \in Q \,\}$ by definition.

PROOF. Note first that for every Q-extension $(A,\varphi) \rightarrowtail G' \twoheadrightarrow Q$ one has $A/[A,G'] = A_Q$. By 2.1 and 1.5 there is a morphism $(\psi \circ ab, \cdot, 1): e \to \psi\, ab(e)$ of extensions. The naturality of θ_* , when applied to this morphism, yields $\theta_*[\psi\, ab(e)] = \psi_c\theta_*(e)$. □

Recall that Cext(Q,A) denotes the central and Ext(Q,A) the abelian extensions of the abelian group A by the group Q.

3.8 UNIVERSAL COEFFICIENT THEOREM. There is a natural short-exact sequence

$$(3.6)\qquad \mathrm{Ext}(Q_{ab},A) \overset{\Psi}{\rightarrowtail} \mathrm{Cext}(Q,A) \overset{\theta_*}{\twoheadrightarrow} \mathrm{Hom}(M(Q),A)\ ,$$

which is split. Here $\Psi[e] = ab^*[e]$ for $ab\colon Q \to Q_{ab}$ the natural map, and $\theta_* = \{[e] \longmapsto \theta_*(e)\}$ is defined by Proposition 3.5.

PROOF. a) Claim: Ψ is a monomorphism. By Theorem 2.4, Ψ is a homomorphism. Let $e\colon A \rightarrowtail E \overset{\sigma}{\twoheadrightarrow} Q_{ab}$ be an abelian extension such that $ab^*[e]$ splits. Then Proposition 1.12 gives a lifting $s\colon Q \to E$ with $\sigma s = ab$. Since E is abelian, s factors as $s = \tau \circ ab$. Thus e is split by $\tau\colon Q_{ab} \to E$.

b) Claim: θ_* is a split epimorphism. Fix a free presentation $e_o\colon R \hookrightarrow F \twoheadrightarrow Q$ of Q. Since a subgroup B of a free abelian group is again free abelian and since $M(F) = 0$ by 3.4, Proposition 3.5 yields an exact sequence

$$(3.7)\qquad 0 \to M(Q) \xrightarrow{\theta_*(e_o)} R/[R,F] \longrightarrow B \to 0$$

with $B = \mathrm{Im}\,\varkappa'$ free abelian. Thus (3.7) is split by some homomorphism $t\colon R/[R,F] \to M(Q)$ with $t \circ \theta_*(e_o) = 1$. The natural map $\pi\colon R_{ab} \twoheadrightarrow R/[R,F]$ induces an isomorphism $\pi^*\colon \mathrm{Hom}(R/[R,F],A) \to$ $\to \mathrm{Hom}_Q(R_{ab},A)$. Consider the homomorphism $\Sigma = \theta^*(e_o,A)\pi^*t^*$: $\mathrm{Hom}(M(Q),A) \to \mathrm{Cext}(Q,A)$. Then for $\nu\colon M(Q) \to A$ we have $\theta_*\Sigma(\nu) = \theta_*[\theta^*(e_o,A)\nu t\pi] = \nu t\theta_*(e_o) = \nu$ by 3.7, thus $\theta_*\Sigma = 1$. Again by 3.7 we have

$$\theta_*(\psi_* ab(e_o) + \psi'_* ab(e_o)) = (\psi+\psi')\,\theta_*(e_o) =$$
$$= \theta_*(\psi_* ab(e_o)) + \theta_*(\psi'_* ab(e_o))$$

for $\psi, \psi' \in \mathrm{Hom}_Q(R_{ab},A)$. Since $\theta^*(e_o,A)$ is surjective by Lemma 2.2, θ_* is a homomorphism.

c) Claim: $\theta_* \Psi = 0$. Let $e: A \rightarrowtail E \twoheadrightarrow Q_{ab}$ be an abelian extension. The naturality of (3.3), when applied to $(1,\cdot,ab) : (e)ab \to e$, gives $\theta_*(\Psi[e]) = \theta_*(e)\, M(ab)$. Now $\operatorname{Im} \theta_*(e) = \operatorname{Ker}(A \to E_{ab}) = 0$ since E is abelian. Thus $\theta_* \Psi[e] = 0$.

d) Claim: $\operatorname{Ker} \theta_* \subseteq \operatorname{Im} \Psi$. Given $[e] \in \operatorname{Cext}(Q,A)$ with $\theta_*(e) = 0$. Clearly the diagram

$$\begin{array}{rcccccc} e: & A & \overset{\kappa}{\rightarrowtail} & G & \overset{\pi}{\twoheadrightarrow} & Q \\ & \| & & \downarrow{\scriptstyle ab} & & \downarrow{\scriptstyle ab} \\ e_1: & A & \overset{\kappa'}{\longrightarrow} & G_{ab} & \overset{\pi_{ab}}{\longrightarrow} & Q_{ab} \end{array}$$

is commutative. The bottom row is short exact since $\theta_*(e) = 0$ in the exact sequence (3.3). Thus $[e] = ab^*[e_1]$ by 1.10.

e) Ψ and θ_* are natural with respect to homomorphisms $\gamma: Q' \to Q$ and $\alpha: A \to A'$, i.e. $\alpha_* \gamma^*(ab^* e) = ab^* \alpha_* \gamma^*_{ab}\, e$ for $e \in \operatorname{Ext}(Q_{ab},A)$ and $\theta_*(\alpha_* \gamma^* e) = \alpha\, \theta_*(e)\, M(\gamma)$ for $e \in \operatorname{Cext}(Q,A)$. Concerning Ψ , this follows from $\gamma_{ab} \circ ab = ab \circ \gamma$ and 1.11. The naturality of θ_* has been explained after Proposition 3.5. □

3.9 COROLLARY. Let Q and A be abelian groups. All central extensions of A by Q are even abelian if, and only if, $\operatorname{Hom}(M(Q),A) = 0$. If every central extension by the abelian group Q is abelian, then $M(Q) = 0$.

PROOF. As Q is abelian, let us identify Q with Q_{ab} by $ab: Q \cong Q_{ab}$. Then the map Ψ of the theorem is just regarding an abelian extension as a central extension. Thus all central extensions of A by Q are abelian precisely when Ψ is surjective, the latter is by Theorem 3.8 equivalent to $\theta_* = 0$. The second assertion follows from the first one, case $A := M(Q)$. For then $\operatorname{Hom}(M(Q),M(Q)) = 0$ implies $M(Q) = 0$. □

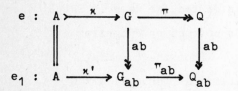

3.10 REMARKS. (a) Fix the group Q , but vary the abelian group A . Then $\mathrm{Coker}\,\Psi$ is a functor from the category of abelian groups into itself. Theorem 3.8 implies that this functor has a representing object, viz. $M(Q)$. This feature could be used for a coordinate-free definition of $M(Q)$ but would be rather abstract.

(b) We noticed already in Corollary 3.9 that the case of abelian Q is quite interesting. For this particular case, BAER [2] obtained Theorem 3.8 as follows. Suppose given a central extension $e: A \overset{\varkappa}{\rightarrowtail} G \overset{\pi}{\twoheadrightarrow} Q$ with abelian Q . Then the commutator $[\ ,\ ]$ in G determines an alternating bilinear form $\phi: Q \times Q \to A$ by $\phi(\pi g_1, \pi g_2) = \varkappa^{-1}[g_1, g_2]$ resp. a homomorphism $\phi: Q \wedge Q \to A$ where $Q \wedge Q$ denotes the <u>exterior square</u> $Q \otimes Q / \langle q \otimes q \mid q \in Q \rangle$. Now the "commutator form" ϕ vanishes precisely when e is an abelian extension. Using factor systems, BAER [2] showed that $\gamma = \{ e \longmapsto \phi \}$ is a homomorphism such that the following sequence is exact:

$$(3.8) \qquad 0 \longrightarrow \mathrm{Ext}(Q,A) \longrightarrow \mathrm{Cext}(Q,A) \xrightarrow{\gamma} \mathrm{Hom}(Q \wedge Q, A) \longrightarrow 0 .$$

Comparison implies a theorem by C. MILLER [1], viz. $M(Q) \cong Q \wedge Q$ for abelian groups Q . We will give a direct proof of Miller's theorem in 4.7 below and show in 4.8 that Baer's commutator form is an interpretation of $\theta_*(e)$. Baer's results are readily accessible in WARFIELD [1; §5].

(c) Let us call a group Q <u>absolutely abelian</u>, if every central extension by Q is abelian. It is a standard fact that every cyclic group is absolutely abelian. Corollary 3.9 implies: All abelian groups Q with trivial multiplicator are absolutely abelian. Now a finite abelian group has trivial multiplicator precisely when it is cyclic. But there are plenty of such infinite abelian groups, e.g. the rationals $\mathbb{Q}$ and all divisible torsion groups. On the other hand, MOSKALENKO [1] has determined the absolutely abelian groups without appeal to the multiplicator, basing himself on (3.8) instead. This theme of absolutely abelian groups will be resumed in Examples IV.4.8 and IV.6.17(b).

3.11 EXAMPLE. We show that the splitting in the Universal Coefficient Theorem is not natural. Let $Q = \mathbb{Z}/4 \rtimes \mathbb{Z}/2$ be the dihedral group of order 8, $U \cong \mathbb{Z}/2$ its commutator subgroup, and $i: U \hookrightarrow Q$ the inclusion map. Then Theorem 3.8 with $A = \mathbb{Z}/2$ gives a commutative diagram with split-exact rows:

$$\begin{array}{ccccc} \mathrm{Ext}(U,\mathbb{Z}/2) & \rightarrowtail & \mathrm{Cext}(U,\mathbb{Z}/2) & \twoheadrightarrow & \mathrm{Hom}(M(U),\mathbb{Z}/2) \\ {\scriptstyle \mathrm{Ext}(i_{ab},1)}\uparrow & & \uparrow{\scriptstyle \mathrm{res}} & & \uparrow{\scriptstyle M(i)^*} \\ \mathrm{Ext}(Q_{ab},\mathbb{Z}/2) & \rightarrowtail & \mathrm{Cext}(Q,\mathbb{Z}/2) & \twoheadrightarrow & \mathrm{Hom}(M(Q),\mathbb{Z}/2) \end{array} .$$

Now $M(U) = 0$ since U is cyclic, thus $M(i)^* = 0$. Moreover, $i_{ab}: U_{ab} \to Q_{ab}$ vanishes by the choice of U , hence $\mathrm{Ext}(i_{ab},1) = 0$. If Theorem 3.8 admitted a natural splitting- or at least some splitting compatible with $i: U \to Q$, then the middle terms Cext would allow direct sum decompositions such that the middle map $\mathrm{res} = i^*$ becomes the direct sum of the maps on the sides, hence res would also vanish. We shall exhibit a central extension $e \in \mathrm{Cext}(Q,A)$ such that its restriction $e|U$ does not split; this means $\mathrm{res} \neq 0$. To this end, let $G = \mathbb{Z}/8 \rtimes \mathbb{Z}/2$ be the dihedral group of order 16 and $\pi: G \longrightarrow Q$ be induced by an epimorphism $\mathbb{Z}/8 \to \mathbb{Z}/4$. Then $\pi^{-1}(U) = [G,G]$ is cyclic of order four. Thus the diagram

$$\begin{array}{rccccc} e/U : & \mathbb{Z}/2 & \hookrightarrow & \pi^{-1}(U) & \twoheadrightarrow & U \\ & \| & & \downarrow & & \downarrow{\scriptstyle i} \\ e : & \mathrm{Ker}\,\pi & \hookrightarrow & G & \xrightarrow{\pi} & Q \end{array}$$

exhibits $\mathrm{res}[e] = [e|U] \neq 0$ by 1.8.

3.12 DEFINITION. An extension e as in (1.1) is called a <u>stem extension</u>, if it is central and $\varkappa N \subseteq [G,G]$.

Proposition 3.5 implies that a central extension e is stem precisely when $\theta_*(e)$ is surjective. Stem extensions will repeatedly occur in the sequel because they are very useful for the computation of Schur multiplicators.

3.13 PROPOSITION. Let Q be a group and N an abelian group. Every central extension class of N by Q is induced forward from a stem extension (with factor Q) if, and only if, $\mathrm{Ext}(Q_{ab},N) = 0$.

PROOF. (a) Assume $\mathrm{Ext}(Q_{ab},N) = 0$. Pick any central extension class $e: N \rightarrowtail G \twoheadrightarrow Q$. By 3.5, $\theta_*(e): M(Q) \to N$ factors as $\theta_*(e) = i \circ \tau$ with $\tau: M(Q) \twoheadrightarrow N \cap [G,G] = N_1$ an epimorphism and $i: N_1 \subseteq N$ an inclusion. Let $e_1 \in \mathrm{Cext}(Q,N_1)$ be any extension class with $\theta_*(e_1) = \tau$. Such e_1 exists by Theorem 3.8 and is a stem extension. By the naturality of θ_* , when applied to $e_1 \to i_* e_1$, we have $\theta_*(i_* e_1) = i\tau = \theta_*(e)$. We now invoke 3.8 for $A := N$. Since $\mathrm{Ext}(Q_{ab},N) = 0$, the map θ_* is isomorphic and hence $e = i_* e_1$ as desired.

(b) Assume that every central extension of N by Q is induced by a stem extension. Then for an arbitrary $e \in \mathrm{Ext}(Q_{ab},N)$, we find a morphism $(j,\cdot,1): e_1 \to \Psi(e)$ of extensions, where e_1 is stem. The naturality of θ_* , when applied to $e_1 \to \Psi(e)$, gives $j\,\theta_*(e_1) = \theta_*\Psi(e_1) = 0$. Since $\theta_*(e_1)$ is surjective, we conclude $j = 0$. Thus $e = \Psi^{-1} j_*(e_1) = 0$. □

3.14 PROPOSITION. (a) Every inner automorphism of G induces the identity of $M(G)$.

(b) Let U be a subgroup of the group G and N the normalizer of U in G , let $i: U \to G$ be the inclusion map. Then the kernel of $res = M(i): M(U) \to M(G)$ contains

$$\{ {}^{n}m \cdot m^{-1} \mid m \in M(U),\ n \in N \} ;$$

in other words, $M(i)$ factors over $M(U) \to M(U)_N$. Here the action of N on $M(U)$ is induced by conjugation in G , restricted to U .

PROOF. (a) Let $M(G)$ be evaluated at $e: R \hookrightarrow F \twoheadrightarrow G$. Any inner automorphism γ of G can be lifted to an inner automorphism of F , say to conjugation by $x \in F$. Now $M(\gamma)$ maps the typical element $r[R,F]$ of $M(G)_e$ onto $xrx^{-1}[R,F] = r[R,F]$.

(b) Let $n \in N$ determine the automorphisms α of U and β of G by conjugation in G . Then the diagram

$$\begin{array}{ccc} M(U) & \xrightarrow{M(i)} & M(G) \\ {\scriptstyle M(\alpha)}\downarrow & & \downarrow{\scriptstyle M(\beta)} \\ M(U) & \xrightarrow{M(i)} & M(G) \end{array}$$

is commutative by the functor property of M . Now $M(\beta) = 1$ by the previous step. Consequently, $res({}^{n}m) = res \circ M(\alpha)\, m = res(m)$ for $m \in M(U)$ and thus ${}^{n}m \cdot m^{-1} \in Ker(res)$. □

4. The Ganea Map of Central Extensions

In the case of a central group extension, the 5-term exact sequence can be extended one step further by the Ganea map. This enlarged sequence has many uses later. An immediate application is a theorem of C. MILLER [1] which describes $M(A)$ as the second exterior power $A \wedge A$ for abelian groups A. Our description of the Ganea map follows ECKMANN/HILTON/STAMMBACH [1]; it is elementary and rests on various commutator identities (mainly due to P. Hall).

Recall ${}^{x}y = xyx^{-1}$ and $[x,y] = xyx^{-1}y^{-1} = {}^{x}y \cdot y^{-1}$.

4.1 LEMMA. (a) $[xy,z] = {}^{x}[y,z] \cdot [x,z]$,

(b) $[x,yz] = [x,y] \cdot {}^{y}[x,z] = [x,y] \cdot [x,z] \cdot [[z,x],y]$,

(c) $[y,x]^{-1} = [x,y] = {}^{xy}[x^{-1},y^{-1}]$,

(d) $[{}^{z}x,[y,z]] \cdot [{}^{y}z,[x,y]] \cdot [{}^{x}y,[z,x]] = 1$,

(e) ${}^{x}[z,[x^{-1},y]] \cdot {}^{y}[x,[y^{-1},z]] \cdot {}^{z}[y,[z^{-1},x]] = 1$. □

4.2 LEMMA. Let A and B be subgroups of the group G .

(a) If G is generated by A and B , then $[A,B]$ is normal in G .

(b) If A and B are generated by the sets S and T , respectively, then the normal closure of $[A,B]$ in G is generated by $\{ [s,t] \mid s \in S, t \in T \}$ as a normal subgroup.

PROOF. Both assertions easily follow from Lemma 4.1 (a), (b). In case (a), it suffices to show that $[A,B]$ is closed under conjugation by elements $g \in A$ or $g \in B$. In case (b), one also needs $[s^{-1},z] = s^{-1}[z,s]s$ and $[x,t^{-1}] = t^{-1}[t,x]t$. □

4.3 THREE-SUBGROUPS LEMMA. Let A,B,C be subgroups and N a normal subgroup of G. If $[A,[B,C]]$ and $[B,[C,A]]$ are contained in N, then so is $[C,[A,B]]$.

PROOF. It is a special case of Lemma 4.2 (b) that $[C,[A,B]] \subseteq N$ provided $[c,[a,b]] \in N$ for all $a \in A$, $b \in B$, $c \in C$. The latter condition follows from the assumptions by 4.1 (e). □

4.4 THEOREM. For every group G with center $Z(G)$ there is a homomorphism

$$(4.1) \qquad \chi_G = \chi(G): G_{ab} \otimes Z(G) \to M(G),$$

called the <u>Ganea map</u>, with the following properties:

(i) If $\gamma: G \to H$ is a group homomorphism with $\gamma Z(G) \subseteq Z(H)$, then $M(\gamma) \circ \chi_G = \chi_H \circ (\gamma_{ab} \otimes \gamma | Z(G))$.

(ii) If $e: N \overset{\varkappa}{\rightarrowtail} G \overset{\pi}{\twoheadrightarrow} Q$ is a central extension and $i: N \to Z(G)$ induced by $\varkappa$, there is an exact sequence

$$(4.2) \qquad G_{ab} \otimes N \xrightarrow{\chi(e)} M(G) \xrightarrow{M(\pi)} M(Q),$$

natural with respect to morphisms of central extensions. Here $\chi(e)$ is defined as the composite

$$G_{ab} \otimes N \xrightarrow{1 \otimes i} G_{ab} \otimes Z(G) \xrightarrow{\chi(G)} M(G).$$

Sequences (3.3') and (4.2) combine to give a longer exact sequence which has first been discovered by GANEA [1] from topological considerations.

PROOF (ECKMANN/HILTON/STAMMBACH [1]). Actually, we first define $\chi(e)$ directly for every central group extension e and show that χ is natural with respect to morphisms. We then define $\chi(G) = \chi(e_G)$, where $e_G: Z(G) \hookrightarrow G \overset{nat}{\twoheadrightarrow} G/Z(G)$ is the obvious extension. Thus (i) and the formula $\chi(e) = \chi(G)(1 \otimes i)$ are immediate.

We again use diagram (3.4) as a "free presentation" of e. Now $[R,F] \subseteq S$, since e is central. Hence $\operatorname{Ker} M(\pi) = [R,F]/[S,F]$. Define a set map $c\colon F \times R \to [R,F]/[S,F]$ by $c(f,r) = [f,r]\cdot[S,F]$. Since $[F,[F,R]] \subseteq [F,S]$, the image of c is central in $F/[S,F]$. Therefore the commutator identities 4.1 (a),(b) become $c(f_1\cdot f_2,r) = c(f_1,r)\cdot c(f_2,r)$ and $c(f,r_1\cdot r_2) = c(f,r_1)\cdot c(f,r_2)$. Trivially $c(S \times R) = 1\cdot[S,F] = c(F \times S)$. Moreover $c([F,F] \times R) = 1\cdot[S,F]$ since c is a homomorphism in the first variable for each fixed $r \in R$. All told, c determines a bilinear map

$$G_{ab} \times N \cong F/([F,F]S) \times R/S \longrightarrow [R,F]/[S,F] \, .$$

The induced map $G_{ab} \otimes N \to [R,F]/[S,F]$ is certainly surjective. Define $\chi(e)$ as the composite $G_{ab} \otimes N \twoheadrightarrow [R,F]/[S,F] \hookrightarrow M(Q)_{e''}$, so that (4.2) is exact. Thus χ is given by

$$(4.3) \qquad \chi(g[G,G] \otimes n) = [f,r]\cdot[S,F] \in M(G)_{e''} \, , \; f \in \rho^{-1}(g), \; r \in \sigma^{-1}(n) .$$

If $(\cdot,\cdot,\gamma)\colon e \to \bar{e}$ is a morphism of extensions and $(3.\bar{4})$ a free presentation of $\bar{e}$, then there exists some translation of diagrams $(3.4) \to (3.\bar{4})$ over $(\cdot,\cdot,\gamma)$, whence $\bar{\chi} = M(\gamma)_{e''|\bar{e}''} \circ \chi$. From the special case $e = \bar{e}$ and $\gamma = 1$, we conclude that χ is independent from the coordinate system e''. □

4.5 Suppose given an abelian group A and apply the preceding theorem to it. This yields a homomorphism $\chi(A)\colon A \otimes A \to M(A)$, which is natural with respect to homomorphisms $\gamma\colon A \to B$ of abelian groups, i.e. the following diagram is commutative

$$\begin{array}{ccc} A \otimes A & \xrightarrow{\chi(A)} & M(A) \\ {\scriptstyle \gamma \otimes \gamma}\downarrow & & \downarrow{\scriptstyle M(\gamma)} \\ B \otimes B & \xrightarrow{\chi(B)} & M(B) \end{array} \, .$$

Considering (4.2) for the degenerate extension $A = A \to 0$, we conclude that $\chi = \chi(A)$ is an epimorphism. If $M(A)$ is evaluated at the free presentation $R \hookrightarrow F \overset{\pi}{\twoheadrightarrow} A$, then $\chi(\pi f \otimes \pi g) = [f,g]\cdot[R,F]$ for $f,g \in F$. In particular, $\chi(a \otimes a) = 0$ for all $a \in A$. Thus χ induces a natural epimorphism $\chi_o: A \wedge A \to M(A)$, where $A \wedge A = \Lambda^2 A$ denotes the exterior square $(A \otimes A)/\langle a \otimes a \mid a \in A\rangle$. Our aim is to show that χ_o is an isomorphism. Note: We write $A \otimes B$ and $\Lambda^2(A)$ additively, but $M(A)$ multiplicatively.

4.6 EXAMPLES. If A is cyclic, then $\Lambda^2 A \cong M(A) \cong 0$. We now compute the Schur multiplicator of any abelian group on two generators by considering suitable stem extensions.

Results: (i) $M(\mathbb{Z} \times \mathbb{Z}) \cong \mathbb{Z}$,

(ii) $M(\mathbb{Z}/m \times \mathbb{Z}/n) \cong \mathbb{Z}/n$ whenever $n \mid m$, and $m = o$ is allowed. In both cases χ_o is an isomorphism.

ad(i). In the group $SL(3,\mathbb{Z})$ of integral 3×3-matrices with determinant One consider the subset

$$T = \{ T(j,k,l) = \begin{pmatrix} 1 & k & l \\ 0 & 1 & j \\ 0 & 0 & 1 \end{pmatrix} \mid j,k,l \in \mathbb{Z} \} .$$

This subset is actually a subgroup with the multiplication rule

(4.4) $\quad T(j,k,l)\cdot T(j',k',l') = T(j+j',k+k',l+l'+kj')$;

thus $T(j,k,l)^{-1} = T(-j,-k,-l+kj)$ and the commutators are

(4.5) $\quad [T(j',k',l'),T(j,k,l)] = T(0,0,k'j-j'k)$.

The assignment $T(j,k,l) \longmapsto (j,k)$ obviously defines a homomorphism of T onto $\mathbb{Z} \times \mathbb{Z}$, with kernel $[T,T] = \{ T(0,0,l) \mid l \in \mathbb{Z} \}$. Moreover, $[T,T]$ is central in T . Applying Proposition 3.5 to the extension $e: [T,T] \hookrightarrow T \twoheadrightarrow \mathbb{Z} \times \mathbb{Z}$, we obtain an epimorphism $M(\mathbb{Z} \times \mathbb{Z}) \twoheadrightarrow [T,T] \cong \mathbb{Z}$. As $\Lambda^2(\mathbb{Z} \times \mathbb{Z}) \cong \mathbb{Z}$, we have another epimorphism $\chi_o: \mathbb{Z} \cong \Lambda^2(\mathbb{Z} \times \mathbb{Z}) \to M(\mathbb{Z} \times \mathbb{Z})$. Since every epi-endomorphism of $\mathbb{Z}$ is an isomorphism, χ_o is an isomorphism. This also implies that $\theta_*(e): M(\mathbb{Z} \times \mathbb{Z}) \to [T,T]$ is an isomorphism.

ad (ii). Assume $n|m$. The subset $K = \{ T(\lambda m,\mu n,\nu n) \mid \lambda,\mu,\nu \in \mathbb{Z} \}$ is a normal subgroup of T , let $T' = T/K$. The assignment $T(j,k,l) \longmapsto (j+m\mathbb{Z},k+n\mathbb{Z})$ induces an epimorphism of T' onto $\mathbb{Z}/m \times \mathbb{Z}/n$, the kernel of which is

$$([T,T]\cdot K)/K \cong [T,T]/([T,T] \cap K) \cong \mathbb{Z}/n \ .$$

Now $\Lambda^2(\mathbb{Z}/m \times \mathbb{Z}/n) \cong \mathbb{Z}/n$. Applying Proposition 3.5 to the central extension $e'\colon ([T,T]\cdot K)/K \rightarrowtail T' \twoheadrightarrow \mathbb{Z}/m \times \mathbb{Z}/n$, we obtain an epi-endomorphism $\mathbb{Z}/n \cong \Lambda^2(\mathbb{Z}/m \times \mathbb{Z}/n) \twoheadrightarrow M(\mathbb{Z}/m \times \mathbb{Z}/n) \twoheadrightarrow \mathbb{Z}/n$. Hence χ_o and $\theta_*(e')$ are isomorphisms.

4.7 THEOREM (C.MILLER [1;Thm.3]). For abelian groups A the natural map $\chi_o\colon A \wedge A \to M(A)$, as defined in 4.5, is an isomorphism.

PROOF. Our proof is elementary, though a direct limit argument can be recognized. Let $x = \sum_{i=1}^{r} (a_i \wedge b_i)$ in $\Lambda^2(A)$ with $\chi_o(A)x = 1$ in $M(A)$. We wish to show $x = 0$. Let $e\colon R \hookrightarrow F \overset{\rho}{\twoheadrightarrow} A$ be a free presentation of A . Choose $g_i,h_i \in F$ such that $\rho(g_i) = a_i$ and $\rho(h_i) = b_i$. Now $\chi_o(A)x = 1$ in $M(A)_e = [F,F]/[R,F]$ means, that there are $r_j \in R$ and $f_j \in F$ with

$$(4.6) \qquad \prod_{i=1}^{r} [g_i,h_i] = \prod_{j=1}^{s} [r_j,f_j]^{\pm 1}$$

in F . Let F_1 be the subgroup of F generated by R and the g_i,h_i,f_j which appear in (4.6). Then $A_1 := \rho(F_1)$ is a finitely generated subgroup of A , which has

$$e_1 : R \hookrightarrow F_1 \xrightarrow{\rho|_{F_1,A_1}} A_1$$

as a free presentation. Let $y = \sum_{i=1}^{r} (a_i \wedge b_i)$ in $\Lambda^2 A_1$ rather than $\Lambda^2 A$. Then $\chi_o(A_1)y = 1$ in $M(A_1)_{e_1}$ since all terms of equation (4.6) lie in F_1 . But clearly $\Lambda^2(i)y = x$, where $i\colon A_1 \hookrightarrow A$ denotes the inclusion. Thus $y = 0$ implies $x = 0$;

the problem is reduced to finitely generated groups A_1 . (See the figure below.)

Now $A_1 \cong \langle t_1 \rangle \times \dots \times \langle t_k \rangle$, a direct product of cyclic groups. We may assume that the order of t_i divides the order of t_j , whenever both orders are finite and $i < j$. If $k = 1$, then $\Lambda^2 A_1 = 0$ and $y = 0$; done. Otherwise $y = \sum_{1 \leq i < j \leq k} \lambda_{ij}(t_i \wedge t_j)$ for suitable integers λ_{ij} . Fix m,n with $1 \leq m < n \leq k$ for the moment. Let $pr: A_1 \twoheadrightarrow A_2 := \langle t_m \rangle \times \langle t_n \rangle$ be the natural projection. Then $z := \Lambda^2(pr)y = \lambda_{mn}(t_m \wedge t_n)$ and $\chi_o(A_2)z = M(pr)\ \chi_o(A_1)y = 1$. (See the figure below.) Since $\chi_o(A_2)$ is an isomorphism by 4.6, we conclude $\lambda_{mn}(t_m \wedge t_n) = 1$ in $\Lambda^2(A_2)$ and hence in $\Lambda^2(A_1)$. Since this holds for all pairs (m,n) , $y = 0$ and $x = 0$ follow.

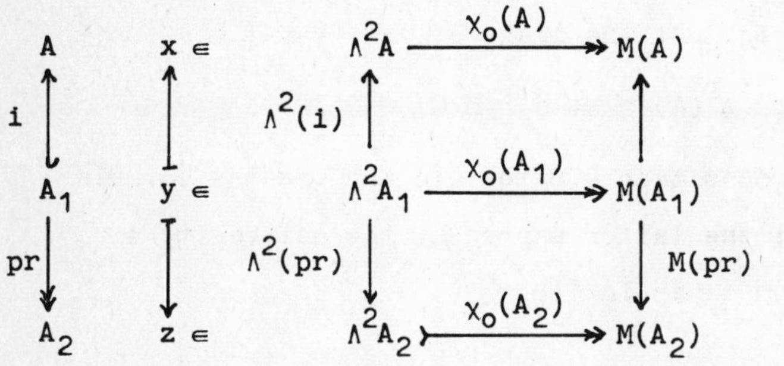

. □

4.8 PROPOSITION. Let $e: N \overset{\varkappa}{\rightarrowtail} G \overset{\pi}{\twoheadrightarrow} Q$ be a central extension with Q abelian, and $\emptyset: \Lambda^2 Q \to N$ be the commutator form of e , as defined in 3.10 (b). Recall the isomorphism $\chi_o(Q): \Lambda^2(Q) \to M(Q)$ from 4.7. Then

$$\emptyset = \theta_*(e) \circ \chi_o(Q) : \Lambda^2 Q \to M(Q) \to N .$$

PROOF. Let $e_o: R \hookrightarrow F \overset{\pi_o}{\twoheadrightarrow} Q$ be the free presentation at which M(Q) shall be evaluated. By 2.2 there is a Q-homomorphism $f: R_{ab} \to (N,0)$ such that $e = f_* ab(e_o)$; thus we have a morphism $(f \circ nat, g, 1): e_o \to e$ of extensions. For any $p,q \in Q$, choose $x,y \in F$ with $\pi_o(x) = p$ and $\pi_o(y) = q$. Then

$\emptyset(p \wedge q) = \varkappa^{-1}[g(x),g(y)] = (f \circ nat)[x,y]$ from the definition of $\emptyset$. On the other hand, by 4.7 and 3.6, $\chi_o(Q)(p \wedge q) = [x,y] \cdot [R,F] \in M(Q)$ and $\theta_*(e)$ is the composite $M(Q) \hookrightarrow R/[R,F] \xrightarrow{f_c} N$. Hence $\theta_*(e)\chi_o(Q)(p \wedge q) = (f \circ nat)[x,y] = \emptyset(p \wedge q)$. □

4.9 REMARK. Consider a central extension $e=(\varkappa,\pi): N \rightarrowtail G \twoheadrightarrow Q$ and assume given $x,y \in Q$ with $xy = yx$. The following construction associates a unique element $z = \emptyset(x,y) \in N$ with such pairs $\{x,y\}$: pick $f,g \in G$ with $\pi(f) = x$ and $\pi(g) = y$, then $[f,g] \in \operatorname{Ker} \pi = \varkappa N$; put $z = \varkappa^{-1}[f,g]$. This ubiquitous construction can be interpreted in terms of Schur multiplicators as follows. Let A be an abelian subgroup containing x and y (e.g. the subgroup generated by x and y) and let $i: A \hookrightarrow Q$ be the inclusion. Then z is just the image of $x \wedge y \in \Lambda^2(A)$ under

$$\theta_*(e)\, M(i)\, \chi_o(A) : \Lambda^2(A) \to M(A) \to M(Q) \to N .$$

For a proof, apply Proposition 4.8 to $e|A$ and use the formula $\theta_*(e|A) = \theta_*(e)M(i)$; the latter expresses the naturality of θ_* with respect to $e|A \to e$ as in 1.8.

5. Compatibility with Other Approaches

We kept our treatment of group extensions and Schur multiplicators as elementary and self-contained as possible; so far we did not treat cohomology and homology groups explicitly. Most of the previous results appear in the literature in terms of (co)homology and this section gives the transition. Since these other formulations make use of resolutions, cocycles, and derived functors, the customary machinery is temporarily needed. This section is rarely used within these notes except for the following section, but will be necessary for most readers who want to do their own research. No originality is claimed.

At first, we are to distinguish three different meanings of the "Schur multiplicator" of a group group Q: $M(Q)$ as defined by the Schur-Hopf formula 3.2, as the second homology group of a certain chain complex, and as the value of the functor $\mathrm{Tor}_2^Q(\mathbb{Z},\mathbb{Z})$ defined by axioms. Likewise, if (A,φ) is a Q-module, we deal with the group $\mathrm{Opext}(Q,A,\varphi)$ of extensions, and the second cohomology group of some cochain complex, and the functor $\mathrm{Ext}_Q^2(\mathbb{Z},(A,\varphi))$. We will exhibit systems of natural isomorphisms which translate our sequences (3.3) and (2.3) into the important (co)homological sequences (5.4) and (5.5). Above all, under these isomorphisms, our formulation 3.8 of the Universal Coefficient Theorem agrees with the one commonly known (in dimension 2).

Here you must keep in mind that the axiomatic (co)homology theory is not unique. For example, assume given a sequence of Ext-functors and connecting homomorphisms ω_n satisfying the axioms. Then you obtain an equivalent theory if you keep the functors Ext but for some n replace all ω_n by $-\omega_n$. Now consider the assertion that an axiomatic map α agrees with some explicit map β after certain identifications. Such a statement can be true only if it is invariant against the indicated sign-changes. Typically the identifications will depend on the connecting homomorphisms, and this relationship must be made explicit before you can prove the assertion $\alpha = \beta$. Because of subtleties like these, a thorough treatment of the translation between the different concepts is necessary.

We are first headed for an interpretation of factor sets in terms of standard free presentations; this has been known for a long time.

5.1 LEMMA. (MAC LANE [1; p.747]). Let Q be a group and $e(Q)$ its standard free presentation as in Definition 3.3. Then R_Q is generated by the elements $J_{p,q} = X_p \cdot X_q \cdot X_{pq}^{-1}$ for $p,q \in Q$; and R_Q is free on $\{ J_{p,q} \mid p,q \in Q , p \neq 1 \neq q \}$.

PROOF. We invoke the Reidemeister-Schreier process in a simple setting, cf. ROTMAN [1; Lemma 11.9, p.244]. We use $S = \{ X_q \mid q \in Q, q \neq 1 \}$ as a basis of F_Q and $T = S \cup \{X_1 = 1\}$ as a system of coset representatives mod R_Q. Clearly T satisfies the Schreier condition. Hence the non-unity elements of $\{ J_{p,q} \mid p,q \in Q , q \neq 1 \}$ form a basis of R_Q. Now $J_{1,q} = 1$ but $J_{p,q} \neq 1$ for $p \neq 1 \neq q$. □

5.2 Let Q be a group and (A,φ) a Q-module. Write F for F_Q and R for R_Q in the standard free presentation $e = e(Q)$ of Q. By Theorem 2.7, e determines an exact sequence, in this special form due to EILENBERG/MAC LANE [2; Thm. 13.1]:

$$\mathrm{Der}(F,A,\varphi\pi) \xrightarrow{\rho(e,A)} \mathrm{Hom}_Q(R_{ab},A) \xrightarrow{\theta^*(e,A)} \mathrm{Opext}(Q,A,\varphi) \longrightarrow 0 .$$

Here R_{ab} is generated by $\bar{J}_{p,q} := J_{p,q} \cdot [R,R]$ and is free abelian on $\{ \bar{J}_{p,q} \mid p,q \in Q ,\ p \neq 1 \neq q \}$. Every homomorphism $f\colon R_{ab} \to A$ satisfies $f_{1,q} = f_{p,1} = 1$ and is Q-linear exactly, if

(5.1) $\quad {}^{p}f_{q,r} \cdot f_{p,q\cdot r} = f_{p,q} \cdot f_{p\cdot q,r}$ for all $p,q,r \in Q$

where $f_{p,q} := f(\bar{J}_{p,q})$. The image of $\rho(e,A)$ consists precisely of those f , for which a set function $g\colon Q \to A$ with $g(1) = 1$ and

(5.2) $\quad f_{p,q} = g(p) \cdot {}^{p}g(q) \cdot g(p\cdot q)^{-1}$ for all $p,q \in Q$

exists. If $f \in \mathrm{Hom}_Q(R_{ab},A)$, then the extension $\theta^*(e,A)f$ has $\{ f_{p,q} \}$ as one of its factor systems.

We notice that conditions (5.1) and (5.2) specify factor sets resp. principal factor sets, as defined in 1.1. By the last assertion, we now obtain the same correspondence between factor sets and extensions as in 1.1. The reader may also consult GRUENBERG [1;§5.3] for the treatment of extensions in terms of arbitrary free presentations.

PROOF. Since F is free, we have $\mathrm{Opext}(F,A,\varphi) = 0$. Now $f\colon R_{ab} \to A$ is Q-linear exactly, if ${}^{p}f(\bar{J}_{q,r}) = f({}^{p}\bar{J}_{q,r}) = f(X_p \cdot J_{q,r} \cdot X_p^{-1} \cdot [R,R])$ for all p,q,r in Q . Condition (5.1) is introduced by the identity

(5.3) $\quad X_p \cdot J_{q,r} \cdot X_p^{-1} = J_{p,q} \cdot J_{p\cdot q,r} \cdot J_{p,q\cdot r}^{-1}$ in $F = F_Q$.

Let d be a derivation of F in $(A,\varphi\pi)$, then $d(J_{p,q}) = d(X_p) \cdot {}^{p}d(X_q) \cdot d(X_{p\cdot q})^{-1}$ and $d(X_1) = 1$. Thus $\rho(e,A)d$ maps $\bar{J}_{p,q}$ onto $g(p) \cdot {}^{p}g(q) \cdot g(p\cdot q)^{-1}$, where $g(r) := d(X_r)$ for r in Q . Conversely, let $g\colon Q \to A$ be a set function with $g(1) = 1$ and let f be defined by (5.2). Since F is free on $\{ X_q \mid q \in Q ,\ q \neq 1 \}$, there is a homomorphism $h\colon F \to (A,\varphi\pi) \lhd F$ with $h(X_q) = (g(q),X_q)$, clearly also $h(X_1) = h(1) = (1,1)$. Thus by 1.9, h defines a derivation d of F in $(A,\varphi\pi)$ with $d(X_q) = g(q)$. Obviously $f = \rho(e,A)d$. Finally, $\{ q \longmapsto X_q \cdot [R,R] \}\colon Q \to F/[R,R]$ is a set section of the projection map in ab(e) and exhibits $\{\bar{J}_{p,q}\}$ as a factor system of ab(e) . If $f \in Hom_Q(R_{ab},A)$, then $\theta^*(e,A)f$ has $\{ f_{p,q} = f(\bar{J}_{p,q}) \}$ as one of its factor systems by 1.5. □

5.3 Let an extension $(\varkappa,\pi)\colon N \rightarrowtail G \twoheadrightarrow Q$ be given, let $(\mathbb{Z},0)$ denote the integers with trivial operators, let A denote the Q-module (A,φ) resp. the G-module $(A,\varphi\pi)$, while H_nQ is short for $H_n(Q,(\mathbb{Z},0))$. Under these assumptions, the (co)homology theory of groups yields the exact sequences

(5.4) $$H_2G \longrightarrow H_2Q \xrightarrow{\beta} N/[N,G] \longrightarrow G_{ab} \longrightarrow Q_{ab} \longrightarrow 0 ,$$

(5.5) $$0 \to Der(Q,A) \longrightarrow Der(G,A) \longrightarrow Hom_Q(N_{ab},A) \xrightarrow{\theta} H^2(Q,A) \longrightarrow H^2(G,A)$$

cf. HILTON/STAMMBACH [1; VI,(8.4) + (8.1)]. These sequences enjoy the naturality properties of 5.4 (b) below.

5.4 REMARK. The following proposition and its proof apply to all functors H_2G and $H^2(G,A)$ satisfying (a) - (c), regardless of the actual constructions. (Method: so-called "abstract nonsense", no chain complexes involved.)

(a) H_2G and $H^2(G,A)$ are covariant functors from the category of groups and the category of copairs (in the sense of 1.15), respectively, to the category of abelian groups.

(b) There are exact sequences (5.4) and (5.5) natural with respect to morphisms of extensions; in addition, (5.5) is natural in the G-module A for fixed G . The unlabelled maps in (5.4) and (5.5) are zero or identity for the degenerate extensions $N = N \longrightarrow 0$ and $0 \rightarrowtail Q = Q$.

(c) Whenever F is a free group and A an F-module, then $H_2F = 0$ and $H^2(F,A) = 0$.

5.5 PROPOSITION. For all groups Q and all Q-modules (A,φ), there are isomorphisms $\xi: M(Q) \to H_2Q$ and $\eta: \mathrm{Opext}(Q,A,\varphi) \to$ $\to H^2(Q,(A,\varphi))$. Here ξ is natural with respect to group homomorphisms and η with respect to copairs. Under the assumptions of 5.3, sequence (3.3) is by $(\xi,\xi,1,1,1)$ translated into (5.4) and sequence (2.3) by $(1,1,1,\eta,\eta)$ into (5.5). The listed properties render ξ and η unique.

PROOF. a) Definition of ξ. Choose any free presentation $e: R \hookrightarrow F \twoheadrightarrow Q$ of Q. Then $M(F) = 0$ by 3.4 and also $H_2F = 0$. Then both $\theta_*(e)$ and $\beta(e)$ are monomorphisms and we define $\xi = \beta(e)^{-1} \circ \theta_*(e)$ as the composite $M(Q) \cong \mathrm{Ker}(R/[R,F] \to F_{ab}) \cong H_2Q$; observe that $R/[R,F] \to F_{ab}$ is uniquely determined by 5.4 (b). Now given free presentations $e_i: R_i \hookrightarrow F_i \longrightarrow Q_i$ for $i=1,2$ and a morphism $(\alpha,\cdot,\gamma): e_1 \to e_2$ of extensions. Since both sequences (3.3) and (5.4) are natural and $\beta(e_2)$ is monomorphic, we conclude $\alpha_* \circ \theta_*(e_1) = \theta_*(e_2) \circ M(\gamma)$ and $\alpha_* \circ \beta(e_1) = \beta(e_2) \circ H_2(\gamma)$ with $\alpha_*: R_1/[R_1,F_1] \to R_2/[R_2,F_2]$, and finally $H_2(\gamma) \circ \xi(e_1) = \xi(e_2) \circ M(\gamma)$. Putting $Q_1 = Q_2$ and $\gamma = 1$, we find that ξ does not depend on the choice of the free presentation and define $\xi(Q)$ as ξ(free presentation). Thus the above formula becomes $H_2(\gamma) \circ \xi(Q_1) =$ $= \xi(Q_2) \circ M(\gamma)$; this is the naturality of ξ.

b) The cohomology case. When F is free then $H^2(F,A) = 0$; moreover $\mathrm{Opext}(F,A) = 0$ since any extension with free quotient group splits. Choose a free presentation e of Q as in a). We then define $\eta(e)$ as the composite $\theta \circ \theta^*(e,A)^{-1}: \mathrm{Opext}(Q,A,\varphi) \cong$ $\cong \mathrm{Coker}(\mathrm{Der}(F,A) \to \mathrm{Hom}_Q(R_{ab},A)) \cong H^2(Q,(A,\varphi))$. It follows as above that $\eta(e)$ does not depend on the choice of e, write $\eta = \eta(Q,A)$ instead. Let $(\gamma,f): (Q_1,A_1) \to (Q_2,A_2)$ be a copair map. Since both sequences (2.3) and (5.5) are natural with respect to copair maps, we obtain $H^2(\gamma,f) \circ \eta(Q_1,A_1) = \eta(Q_2,A_2) \circ \mathrm{Opext}(\gamma,f)$. It follows as in the homology case that η translates sequence (2.3) into sequence (5.5) and is uniquely determined by this property. □

5.6 ADDENDUM. In HILTON/STAMMBACH [1; VI.10], a bijection $\Delta\colon \mathrm{Opext}(Q,A) \to H^2(Q,A)$ is defined as follows. Given $e \in \mathrm{Opext}(Q,A)$, consider the exact sequence (5.5) for e and the Q-module A; put $\Delta(e) = \theta(1_A) \in H^2(Q,A)$. We claim that the map Δ agrees with our map η. Indeed, the definition of $\theta^* = \theta^*(e,A)$ in Theorem 2.7 gives $\theta^*(1_A) = e$. Since $\theta = \eta\circ\theta^*\colon \mathrm{Hom}_Q(A,A) \to H^2(Q,A)$ by 5.5, we have $\Delta(e) = \theta(1_A) = \eta\theta^*(1_A) = \eta(e)$.

5.7 The customary formulation of the Universal Coefficient Theorem (MAC LANE [2; III Theorem 4.1]). Given a group Q and an abelian group A, which is regarded as a Q-module with trivial operators. Then for each $n \in N$ there is a natural short exact sequence

$$(5.6)\qquad \mathrm{Ext}(H_{n-1}Q,A) \overset{\beta}{\rightarrowtail} H^n(Q,A) \overset{\alpha}{\twoheadrightarrow} \mathrm{Hom}(H_nQ,A),$$

where H_nQ denotes integral homology as in 5.3. The maps α and β are defined via chain-complexes as follows. Let $\underline{P}$ be a Q-free resolution of $(\mathbb{Z},0)$. Then $\underline{K} = (\mathbb{Z},0) \otimes_Q \underline{P}$ is a chain-complex of free abelian groups with $H_n(\underline{K}) = H_nQ$ and $H^n(\mathrm{Hom}(\underline{K},A)) = H^n(Q,A)$. Represent $x \in H^n(Q,A)$ by an n-cocycle f and $y \in H_n(Q)$ by an n-cycle c. Then $\alpha(x)$ is the map $\{y \longmapsto f(c)\}$. Pick in $z \in \mathrm{Ext}(H_{n-1}Q,A)$ an abelian extension $e_z\colon A \rightarrowtail E \twoheadrightarrow H_{n-1}Q$. Since the group C_{n-1} of (n-1)-cycles is free-abelian, some dotted maps can be found as to render the following diagram commutative:

$$(5.7)\qquad \begin{array}{ccccccc} & \partial_nK_n & \rightarrowtail & K_{n-1} & \twoheadrightarrow & H_{n-1}Q \\ & {\scriptstyle\varphi}\downarrow & & \downarrow{\scriptstyle\psi} & & \| \\ e_z: & A & \rightarrowtail & E & \twoheadrightarrow & H_{n-1}Q \end{array}$$

Then $\beta(z)$ is the cohomology class of the n-cocycle $\varphi\circ\partial_n'\colon K_n \to \partial_nK_n \to A$.

5.8 PROPOSITION. There are natural isomorphisms $\rho\colon H_1Q \to Q_{ab}$, $\sigma\colon H_2Q \to M(Q)$, and $\tau\colon \mathrm{Opext}(Q,A,\varphi) \to H^2(Q,(A,\varphi))$ for every Q-module (A,φ) such that, in the case of trivial group action on A, the following diagram with (3.6) and (5.6) is commutative:

$$\begin{array}{ccccc} \mathrm{Ext}(Q_{ab},A) & \xrightarrow{\Psi} & \mathrm{Cext}(Q,A) & \xrightarrow{\theta_*} & \mathrm{Hom}(M(Q),A) \\ \cong\downarrow\rho^* & & \cong\downarrow\tau & & \cong\downarrow\mathrm{Hom}(\sigma,A) \\ \mathrm{Ext}(H_1Q,A) & \xrightarrow{\beta} & H^2(Q,A) & \xrightarrow{\alpha} & \mathrm{Hom}(H_2Q,A) \end{array} .$$

This proposition provides a conceptual description of α and β . The interpretation of α seems to be "folklore", see also ECKMANN/HILTON/STAMMBACH [1; Thm. 2.2].

PROOF. We evaluate $M(Q)$ at the standard free presentation $e=e(Q)\colon R \hookrightarrow F \twoheadrightarrow Q$ (notation as in 5.1 and 5.2). We compute $H^2(Q,A)$ and H_1Q and H_2Q at the normalized inhomogeneous bar resolution $\underline{B}(Q)$ resp. at $\underline{K}(Q) = (\mathbb{Z},0) \otimes_Q \underline{B}(Q)$. The relevant part of $\underline{K} = \underline{K}(Q)$ is

$$\cdots \longrightarrow K_3 \xrightarrow{\partial_3} K_2 \xrightarrow{\partial_2} K_1 \xrightarrow{\partial_1} \mathbb{Z} \longrightarrow 0 .$$

Here K_i for $i = 1,2,3$ is the free abelian group on all i-tuples $[x_1|\dots|x_i]$, where the x_j run through all non-unity group elements; let $[x_1|\dots|x_i] = 0$ whenever some $x_j = 1$. Then the boundary operators are given by $\partial_1 = 0$ and

(5.8.1) $\qquad \partial_2[x|y] = [x] + [y] - [x\cdot y]$,

(5.8.2) $\qquad \partial_3[x|y|z] = [y|z]+[x|y\cdot z] - [x\cdot y|z] - [x|y]$.

a) The definition of ρ , cf. MAC LANE [2; p.290]. The homomorphism $|[x] \longmapsto x\cdot[Q,Q]|\colon K_1=\mathrm{Ker}(\partial_1) \to Q_{ab}$ sends ∂_2K_2 to 0 and thus defines $\rho\colon H_1Q \to Q_{ab}$. It is not difficult to construct a homomorphism $Q_{ab} \to H_1Q$ inverse to ρ . If $f\colon Q \to Q'$ is a group homomorphism, there is an obvious induced chain map

$\underline{K}(f)\colon \underline{K}(Q) \to \underline{K}(Q')$. Inspection gives that ρ is compatible with the pair $\{K_1(f)\ ,\ f_{ab}\colon Q_{ab} \to Q'_{ab}\}$.

b) The definition of σ . Note that $\operatorname{Ker} \partial_2$ consists precisely of those elements $\sum n_{x,y}[x|y]$ for which

(5.9) $\sum_{y\in Q}(n_{x,y} + n_{y,x} - n_{xy^{-1},y})$ for all $x \in Q\setminus\{1\}$.

(For convenience, we allow $y = 1$ etc., but put $n_{x,1} = 0 = n_{1,y}$.) By 5.2 a typical element of R_{ab} is $b\cdot[R,R]$ for some $b = \Pi\,(J_{x,y})^{n_{x,y}}$. Now $b\cdot[R,F]$ lies in $M(Q) = \operatorname{Ker}(R/[R,F] \to F_{ab})$ exactly, if $R \longrightarrow F \longrightarrow F_{ab}$ sends b to 1 . Since $J_{x,y} = X_x\cdot X_y\cdot X_{xy}^{-1}$ in F , this condition is equivalent to (5.9). Clearly $\bar\sigma = \{[x|y] \longmapsto \bar J_{x,y}\}\colon K_2 \to R_{ab}$ defines an isomorphism (between free-abelian groups). The above argument gives $\bar\sigma(\operatorname{Ker}\partial_2) = (R \cap [F,F])/[R,R]$. Comparing (5.8.2) with (5.3), we find

$$\bar\sigma\, \partial[x|y|z] = [J_{y,z},X_x]\cdot[R,R]$$

and conclude $\bar\sigma(\partial_3 K_3) = [R,F]/[R,R]$. Hence $\bar\sigma$ induces an isomorphism $\sigma\colon H_2Q = (\operatorname{Ker}\partial_2)/\partial_3K_3 \to M(Q) = (R \cap [F,F])/[R,F]$. The argument for the naturality of σ is similar to that used for ρ .

c) The definition of τ . Here we compute $H^2(Q,(A,\varphi))$ at $\underline{B}(Q)$. By 2.2 every extension class $e_1 \in \operatorname{Opext}(Q,A,\varphi)$ can be written as $e_1 = \theta^*(e,A)f$ for some $f \in \operatorname{Hom}_Q(R_{ab},A)$. Now 5.2 implies that $\{[x|y] \longmapsto f_{x,y}\} = f(\bar J_{x,y})\colon B_2(Q) \to A$ is a 2-cocycle which is uniquely determined modulo 2-coboundaries. Let τ be the resulting isomorphism $\{e_1 \longmapsto \{f_{x,y}\}\}\colon \operatorname{Opext}(Q,A,\varphi) \to \operatorname{Ker}(\partial^2)/\operatorname{Im}(\partial^1) = H^2(Q,(A,\varphi))$. Since every group homomorphism $f\colon Q \to Q'$ induces an obvious chain map $\underline{B}(f)\colon \underline{B}(Q) \to \underline{B}(Q')$, the proof for the naturality of τ with respect to copair maps is straightforward.

d) Claim: $\tau\circ\Psi = \beta\circ\rho^*$. Given $e_2 \in \operatorname{Ext}(Q_{ab},A)$, then $\Psi(e_2) = ab^*(e_2)$ with $ab\colon Q \longrightarrow Q_{ab}$ the natural map. By Lemma 2.2 we find a Q-linear map $f\colon R_{ab} \to A$ such that $ab^*(e_2) = f_*[ab(e)]$, and we have a commutative diagram

$$\begin{array}{rcccccc}
ab(e): & R_{ab} & \rightarrowtail & F/[R,R] & \twoheadrightarrow & Q \\
 & \downarrow{\scriptstyle f} & & \downarrow{\scriptstyle g} & & \| \\
ab^*(e_2): & A & \rightarrowtail & \tilde{E} & \twoheadrightarrow & Q \\
 & \| & & \downarrow{\scriptstyle h} & & \downarrow{\scriptstyle ab} \\
e_2: & A & \overset{\iota}{\rightarrowtail} & E & \overset{\pi}{\twoheadrightarrow} & Q_{ab}
\end{array} .$$

Hence $\tau \circ \Psi(e_2)$ is represented by the 2-cocycle $\{f_{x,y} = f(\bar{J}_{x,y})\}$. It is immediate that $\rho^*(e)$ can be described as $A \overset{\iota}{\rightarrowtail} E \overset{\pi'}{\twoheadrightarrow} H_1Q$ where $\pi' = \rho^{-1}\pi$. Define $\psi\colon K_1 = \mathrm{Ker}(\partial_1) \to E$ by $\psi[x] = hg(X_x \cdot [R,R])$. Then $\xi\pi'\psi[X_x] = x[Q,Q]$, thus ψ fits into diagram (5.7) and yields φ by restriction. Now $\beta \circ \rho^*(e)$ is represented by the 2-cocycle $\varphi\partial_2\colon K_2 \to A$ which maps $[x|y]$ onto $\iota^{-1}hg(X_x \cdot X_y \cdot X_{xy}^{-1}) = f(\bar{J}_{x,y})$.

e) Claim: $\alpha \circ \tau = \mathrm{Hom}(\sigma,A) \circ \theta_*$. Given $e_1 \in \mathrm{Cext}(Q,A)$, find $f \in \mathrm{Hom}_Q(R_{ab},A)$ such that $e_1 = f_*[ab(e)]$. Let $c = \sum_{x \neq 1 \neq y} n_{x,y}[x|y]$, with the coefficients satisfying (5.9), be a typical 2-cycle. Then $\tau(e_1)$ is represented by the 2-cocycle $\{f_{x,y}\}$ and $\alpha \circ \tau(e_1)$ is represented by $\{c \longmapsto \Sigma\, n_{x,y}f_{x,y}\}$. On the other hand $\theta_*(e_1) = f_c \circ \theta_*(e)$ by 3.7, while 3.6 exhibits $\mathrm{Hom}(\sigma,A) \circ \theta_*$ as the composite homomorphism

$$H_2Q \xrightarrow{\sigma} M(Q) = (R \cap [F,F])/[F,R] \hookrightarrow R/[R,F] \xrightarrow{f_c} A_Q = A ,$$

which maps the homology class of c also onto $\sum n_{x,y}f_{x,y}$. (Here we employed additive rather than multiplicative notation for the sake of readability.)

5.9 REMARKS. (a) From the proof of Proposition 5.8, we recall the following description of the isomorphism

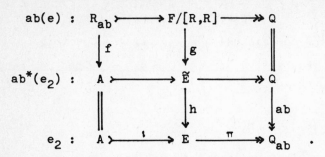

$$\sigma : H_2Q \to M(Q) .$$

Evaluate H_2Q by at the bar resolution and $M(Q)$ at the standard free presentation $e(Q)$. A typical cycle has the form $c = \sum_{x,y\in Q} n_{x,y}[x|y]$, the coefficients $n_{x,y}$ being subject to (5.9) and (without loss of generality) $n_{x,1} = 0 = n_{1,y}$. Then σ maps the homology class of c onto

$$\prod_{x,y\in Q} J_{x,y}^{n_{x,y}}\cdot[R_Q,F_Q] \in M(Q) .$$

(b) Assume given $x,y \in Q$ with $xy = yx$, then the construction of 4.9 for the central extension $c(e(Q))$ yields the element

$$z = \emptyset(x,y) = [X_x,X_y]\cdot[R_Q,F_Q] = J_{x,y}\cdot J_{y,x}^{-1}\cdot[R_Q,F_Q]$$

in $M(Q)$, evaluated at $e(Q)$. In particular, elements generating the image of the Ganea map χ_Q can be obtained in this fashion, cf. (4.3). Now let also $e \in Cext(Q,A)$ with factor system f be given. Then we conclude from Proposition 5.8, combined with the description of σ in (a) and that of α in 5.7:

$$\theta_*(e)\, z = f(x,y)\cdot f(y,x)^{-1} .$$

Finally, when Q is abelian and $\chi_o = \chi_o(Q)$ is as in 4.5, then

$$\theta_*(e)\chi_o(x\wedge y) = f(x,y)\cdot f(y,x)^{-1} \quad \text{for all } x,y \in Q .$$

In a different vein, the function $f(x,y)\cdot f(y,x)^{-1}$ has been studied by IWAHORI/MATSUMOTO [1; §1].

We now can infer results from the (co)homology theory of groups. For example, the direct limit argument asserts that homology (not cohomology) groups of groups respect direct limits also in the group variable. This principle has long been known, a proof is spelled out in BEYL [3]. Of greatest interest to us is the following proposition, in the special case that a group G is regarded as the directed union of its finitely generated subgroups.

5.10 PROPOSITION (Direct Limit Argument). Let $\{ \pi_\alpha^\beta : G_\alpha \to G_\beta \mid \alpha \leq \beta \in I \}$ be a direct system of groups over the directed set I, let G be the direct limit group and $\pi_\alpha : G_\alpha \to G$ the canonical homomorphisms. Then $M(G)$ together with $M(\pi_\alpha): M(G_\alpha) \to M(G)$ is the direct limit of the direct system

$$\{ M(\pi_\alpha^\beta): M(G_\alpha) \to M(G_\beta) \mid \alpha \leq \beta \in I \} .$$

In short, $M(\text{dir.lim. } G_\alpha) = \text{dir.lim. } M(G_\alpha)$. □

6. Corestriction (Transfer)

This section aims at Propositions 6.8 and 6.9, which relate the Schur multiplicator of a finite group G to the order of G and to the multiplicators of the Sylow p-subgroups. The formulation of these results is self-contained and the remainder of this section may be omitted on the first reading.

For the proofs, we use that the functor M is naturally isomorphic to $H_2(-,\mathbb{Z})$ by Proposition 5.8 and invoke the corestriction functor $Cor_2\colon H_2G \to H_2U$, defined whenever $|G:U|$ is finite. The needed properties of corestriction are well-known, though usually stated for cohomology. The fundamental source on corestriction is the paper by ECKMANN [1]. One of his motivating ideas was that Cor^1 is dual to the classical transfer homomorphism, cf. 6.12. Thus we think of $Cor_2\colon M(G) \to M(U)$ as a "higher transfer". We use the opportunity to present an approach which slightly differs from Eckmann's and treats homology and cohomology in the same fashion. (Note that we deal with ordinary co/homology rather than the Tate cohomology of finite groups. In the latter case, a common approach is by dimension shift.)

6.1 ASSUMPTIONS and NOTATION. Let $i\colon U \hookrightarrow G$ be the inclusion of a subgroup U into the group G; choose a right transversal $\{x_k\}$ to U in G, thus

$$(6.1) \qquad G = \bigcup_{k=1}^{m} Ux_k = \bigcup_{k=1}^{m} x_k^{-1}U .$$

(At present, m may be infinite.) Then every left module (A,φ) gives rise to the U-module $V(A,\varphi) = (A,\varphi\circ i)$, what we express as: "A can be considered as a U-module". (Modules are now written additively: 0,+,etc.) The integers Z are given trivial action by G and U.

Obviously V is an exact functor. As (6.1) implies

$$\bigoplus_{k=1}^{m} ZU\cdot x_k = ZG = \bigoplus_{k=1}^{m} x_k^{-1}\cdot ZU ,$$

$V(ZG)$ is U-free as a left as well as a right U-module. Hence $V(P)$ is U-free for every free G-module P.

We regard both

(6.2) $\quad H^n(G,A) \underset{def}{=} Ext_G^n(Z,A)$

and $H^n(U,VA)$ for $n \geq 0$ as functors from the category of left G-modules A, likewise both

(6.3) $\quad H_n(G,B) \underset{def}{=} Tor_n^G(B,Z)$

and $H_n(U,VB)$ as defined for right G-modules B. - Unexplained notation and terminology is that of HILTON/STAMMBACH [1; chp. VI]. We usually identify $H^o(G,A)$ with

(6.4) $\quad A^G \underset{def}{=} \{ a \in A \mid {}^g a = a \text{ for all } g \in G \}$

and $H_o(G,B)$ with $B_G = B/B\cdot IG$ as in 3.7, without explicit mention of the isomorphisms

$$H^o(G,A) \cong Hom_G(Z,A) \cong A^G ,$$

$$H_o(G,B) \cong B \otimes_G Z \cong B_G .$$

In each case, the left-hand isomorphism is induced by the augmentation of any G-free resolution of Z, while the other isomorphism is $\{f \longmapsto f(1)\}$ for $f: Z \to A$ resp. $\{\Sigma(b_t \otimes t) \longmapsto \Sigma(tb_t + B\cdot IG)\}$ for $t \in Z$, $b_t \in B$.

6.2 REMARK. There are isomorphisms

$$\zeta : H_n(U,VB) = \mathrm{Tor}_n^U(VB,\mathbb{Z}) \to \mathrm{Tor}_n^G(B,\mathbb{Z}G \otimes_U \mathbb{Z}) ,$$

$$\zeta : H^n(U,VA) = \mathrm{Ext}_U^n(\mathbb{Z},VA) \to \mathrm{Ext}_G^n(\mathbb{Z}G \otimes_U \mathbb{Z},A)$$

that are natural in the G-modules A and B and for $n = 0$ are described by the formulas

$$\zeta(b + VB\cdot IU) = b \otimes (1 \otimes 1) \in B \otimes_G (\mathbb{Z}G \otimes_U \mathbb{Z})$$

resp. $\zeta(a) = \{g \otimes t \longmapsto {}^g(ta)\} \in \mathrm{Hom}_G(\mathbb{Z}G \otimes_U \mathbb{Z},A)$.

Here $g \in G$ acts on $\mathbb{Z}G \otimes_U \mathbb{Z}$ by ${}^g(a \otimes t) = (gx) \otimes t$ for $x \in \mathbb{Z}G$. (These maps are well-known, cf. HILTON/STAMMBACH [1; Prop. IV.12.2; Lemma VI.6.2] .

Take a G-free resolution $\underline{P} \twoheadrightarrow \mathbb{Z}$ of $\mathbb{Z}$; then $V(\underline{P})$ is a U-free resolution of $\mathbb{Z}$ and $\mathbb{Z}G \otimes_U \underline{P}$ a G-free resolution of $\mathbb{Z}G \otimes_U \mathbb{Z}$ by 6.1. Now ζ^{-1} is induced by the standard isomorphisms

$$B \otimes_G (\mathbb{Z}G \otimes_U V\underline{P}) \cong (B \otimes_G \mathbb{Z}G) \otimes_U V\underline{P} \cong VB \otimes_U V\underline{P} ,$$

$$\mathrm{Hom}_G(\mathbb{Z}G \otimes_U V\underline{P},A) \cong \mathrm{Hom}_U(V\underline{P},\mathrm{Hom}_G(\mathbb{Z}G,A)) \cong \mathrm{Hom}_U(V\underline{P},VA)$$

of chain resp. cochain complexes. This description also implies: The connecting homomorphisms $\omega_n(e)$ of any exact sequence $e: B' \rightarrowtail B \twoheadrightarrow B''$ are respected, i.e. the diagrams

$$\begin{array}{ccc} H_n(U,VB'') & \xrightarrow{\omega_n(e)} & H_{n-1}(U,VB') \\ \downarrow \zeta & & \downarrow \zeta \\ \mathrm{Tor}_n^G(B'',\mathbb{Z}G \otimes_U \mathbb{Z}) & \xrightarrow{\omega_n(e)} & \mathrm{Tor}_{n-1}^G(B',\mathbb{Z}G \otimes_U \mathbb{Z}) \end{array} \tag{6.5}$$

are commutative; similarly in cohomology.

6.3 DEFINITION. With $\epsilon = \{ \sum g \otimes t_g \longmapsto \sum tg \} : \mathbb{Z}G \otimes_U \mathbb{Z} \to \mathbb{Z}$ define <u>restriction</u> as

$$\mathrm{Res}_n = H_n(i,B) = \mathrm{Tor}_n(B,\epsilon)\circ \zeta :$$

$$H_n(U,VB) \to \mathrm{Tor}_n^G(B,\mathbb{Z}G \otimes_U \mathbb{Z}) \to \mathrm{Tor}_n^G(B,\mathbb{Z}) = H_n(G,B) ,$$

$Res^n = H^n(i,A) = \zeta^{-1} \circ Ext^n(\varepsilon,A)$:

$H^n(G,A) = Ext^n_G(Z,A) \to Ext^n_G(ZG \otimes_U Z,A) \to H^n(U,VA)$.

One easily verifies that ε is G-linear and Res^o just is the inclusion $A^G \hookrightarrow (VA)^U$. Moreover, this definition of a restriction map agrees with the "more obvious" ones as in MAC LANE [2; p.116] and HILTON/STAMMBACH [1; p.190].

6.4 LEMMA. Assume $m = |G:U| < \infty$. Then the map

$$\nu_A = \{a \longmapsto \sum_{k=1}^{m} x_k^{-1} \otimes x_k \cdot a\} : A \to ZG \otimes_U A$$

does not depend on the choice of transversal, is G-linear, and natural in the left G-module A. Here $g \in G$ acts on $ZG \otimes_U A$ by ${}^g(x \otimes a) = (gx \otimes a)$ for $x \in ZG$ and $a \in A$.

PROOF. The first assertion follows from

$$(ux_k)^{-1} \otimes (ux_k) \cdot a = x_k^{-1} u^{-1} \otimes ux_k \cdot a = x_k^{-1} \otimes x_k \cdot a$$

for $u \in U$, the naturality is obvious. We are left to show that ν_A is G-linear. Note that $\{x_k g\}$ is again a (right) transversal to U in G , for $g \in G$. Since ν_A is independent of the choice of transversal, we conclude

$$g \cdot \nu_A(a) = g \cdot \sum_{k=1}^{m} (x_k g)^{-1} \otimes (x_k g) \cdot a$$

$$= \sum_{k=1}^{m} x_k^{-1} \otimes x_k \cdot (g \cdot a) = \nu_A({}^g a) \ . \ \square$$

6.5 DEFINITION. Assume $|G:U| < \infty$. With $\nu = \nu_Z : Z \to ZG \otimes_U Z$ define <u>corestriction</u> as

$Cor_n = \zeta^{-1} \circ Tor_n(B,\nu) : H_n(G,B) = Tor_n^G(B,Z) \to Tor_n^G(B,ZG \otimes_U Z) \to H_n(U,VB)$,

$Cor^n = Ext(\nu,A) \circ \zeta : H^n(U,VA) \to Ext^n_G(ZG \otimes_U Z,A) \to Ext^n_G(Z,A) = H^n(G,A)$.

This description puts into evidence that $\{Cor_n\}$ and $\{Cor^n\}$ respect connecting homomorphisms in analogy with (6.5). Tracing the formulas for $Cor_o\colon B_G \to (VB)_U$ and $Cor^o\colon (VA)^U \to A^G$, one finds

(6.6) $$Cor_o(b+B\cdot IG) = \sum_{k=1}^{m} b\cdot x_k^{-1} + VB\cdot IU ,$$

(6.7) $$Cor^o(a) = \sum_{k=1}^{m} x_k^{-1}\cdot a \quad \text{for} \quad a \in (VA)^U .$$

(If, in the context of finite groups, homology groups are identified with Tate cohomology groups of negative degree, then our homology restriction is usually called corestriction and vice-versa.)

6.6 THEOREM. Assume that $m = |G:U|$ is finite. Let m_* denote the multiplication by m in any abelian group. Let A be a left and B a right G-module.

a) $Res_n \circ Cor_n = m_* : H_n(G,B) \to H_n(G,B)$.

b) $Cor^n \circ Res^n = m_* : H^n(G,A) \to H^n(G,A)$.

PROOF. We first invoke the notation of 6.3 and 6.5 and find that $\epsilon\cdot\nu = m_*\colon Z \to Z$. In the homology case, we have

$$Res_n \circ Cor_n = Tor_n(B,\epsilon)\circ\zeta\circ\zeta^{-1}\circ Tor_n(B,\nu)$$
$$= Tor_n(B,\epsilon\circ\nu) = Tor_n(B,m_*) = m_* .$$

In the cohomology case, we likewise conclude

$$Cor^n \circ Res^n = Ext^n(\nu,A)\circ\zeta\circ\zeta^{-1}\circ Ext^n(\epsilon,A)$$
$$= Ext^n(\epsilon\circ\nu,A) = Ext^n(m_*,A) = m_* . \quad \square$$

6.7 COROLLARY. If G is finite of order $|G|$, then $H^n(G,A)$ and $H_n(G,B)$ have exponent dividing $|G|$, for all $n > 0$ and all left G-modules A and all right G-modules B .

PROOF. Apply Theorem 6.6 for $U = 0$, thus $m = |G|$. Note

$H^n(U,VA) = 0 = H_n(U,VB)$ for $n > 0$. Consequently

$$|G|_* = Res_n \circ Cor_n = 0 : \; H_n(G,B) \to H_n(G,B) \text{ , etc.} \quad \square$$

6.8 PROPOSITION. If G is a finite group, then $M(G)$ is a finite abelian group of exponent dividing $|G|$.

PROOF. As the bar resolution of G is finitely generated in degree two, the homology group $H_2(G,\mathbb{Z})$ is finitely generated. By Corollary 6.7, this group is finite of exponent dividing $|G|$. Finally $M(G) \cong H_2(G,\mathbb{Z})$ by Proposition 5.8. $\square$

A representation-theoretic proof of this proposition and a sharper estimate, due to SCHUR [1], will be given with Corollary II.3.10: Roughly, $\sqrt{|G|}$ is an exponent of $M(G)$.

6.9 PROPOSITION. Let G be a finite group and P be a Sylow p-subgroup for some prime p . Then the image of $res: M(P) \to M(G)$ is the Sylow p-subgroup $M(G)_p$ of $M(G)$. Moreover, $M(G)_p$ is isomorphic to a direct summand of $M(P)$.

PROOF. Put $U := P$, then p does not divide $m = |G:P|$. By Proposition 5.8 there is a commutative diagram

$$\begin{array}{ccccc} M(G) & \xrightarrow{\;C\;} & M(P) & \xrightarrow{\;M(i)\;} & M(G) \\ \cong\uparrow\sigma & & \cong\uparrow\sigma & & \cong\uparrow\sigma \\ H_2G & \xrightarrow{\;Cor_2\;} & H_2P & \xrightarrow{\;H_2(i,\mathbb{Z})\;} & H_2G \end{array}$$

where $C := \sigma \circ Cor_2 \circ \sigma^{-1}$. Recall $res = M(i)$ as in Proposition 3.14 (b). We invoke Theorem 6.6 for $n = 2$ to conclude

(6.8) $\quad C \circ M(i) = m_* : \; M(G) \to M(G)$.

Now $M(P)$ is a finite p-group by Proposition 6.8, thus $\operatorname{Im} M(i) \subseteq M(G)_p$ and

$$R \circ C' = m_* : \; M(G)_p \xrightarrow{C'} M(P) \xrightarrow{R} M(G)_p$$

where C' and R are the obvious restrictions of C and $M(i)$, respectively. As $M(G)_p$ is a finite abelian p-group, multiplication by m is an automorphism of $M(G)_p$. Hence R is surjective and $\operatorname{Im} M(i) = M(G)_p$. Moreover, $M(P)$ is the internal direct sum of $\operatorname{Ker} R = \operatorname{Ker} M(i)$ and $\operatorname{Im} C' \cong M(G)_p$. □

6.10 PROPOSITION, cf. JONES/WIEGOLD [1]. If U is a subgroup of finite index m in the group G, then

$$m_*M(G) = \{ x^m \mid x \in M(G) \}$$

is isomorphic to a subquotient of $M(U)$.

Note: If $M(U)$ is a finite group, then any subquotient of $M(U)$ is isomorphic to a subgroup of it.

PROOF. By Theorem 6.6 (together with Proposition 5.8), there is a homomorphism $C\colon M(G) \to M(U)$ such that

$$\operatorname{res} \circ C = m_* : \; M(G) \to M(G) .$$

Hence

$$m_*M(G) = \operatorname{Im}(\operatorname{res} \circ C) \subseteq \operatorname{Im}(\operatorname{res}) \cong M(U)/\operatorname{Ker}(\operatorname{res}) . \quad \square$$

For practical purposes we are going to describe the corestriction in terms of cycles and cocycles, with respect to the (normalized) bar resolution $\underline{B}(G) = \{B_n(G)\}$ on homogeneous generators

(6.9) $\quad (g_0, g_1, \ldots, g_n) = g_0[g_0^{-1} g_1 | g_1^{-1} g_2 | \ldots | g_{n-1}^{-1} g_n]$

and $\underline{B}(U)$. First, the naturality assertion of Lemma 6.4 gives that

$$\{ \nu_{B_n(G)} : B_n(G) \to \mathbb{Z}G \otimes_U B_n(G) \}$$

is a chain transformation lifting $\nu\colon \mathbb{Z} \to \mathbb{Z}G \otimes_U \mathbb{Z}$. Next ζ expresses chains and cochains at the resolution $\underline{B}(G)$, considered as a U-free resolution. Third, $\underline{B}(G)$ is to be compared with $\underline{B}(U)$. While $\underline{B}(i)\colon \underline{B}(U) \to \underline{B}(G)$ is a U-linear lift of 1_Z, a map in the opposite direction is needed. To this end, fix the transversal (6.1) to U in G. Let $\bar{g}$ denote the chosen representative of the coset Ug, then $g = (g\cdot\bar{g}^{-1})\bar{g}$ with $g\cdot(\bar{g})^{-1} \in U$. Due to the boundary formula

$$(6.10) \qquad \partial_n(g_o, g_1, \ldots, g_n) = \sum_{j=1}^{n} (-1)^j (g_o, \ldots, \hat{g}_i, \ldots, g_n)$$

- where the roof means deletion of the term under it, the projection $\pi(g) = g(\bar{g})^{-1}$ onto the U-component induces a U-linear chain transformation $\underline{B}(\pi)\colon \underline{B}(G) \twoheadrightarrow \underline{B}(U)$ over 1_Z. Putting the three steps together, we obtain the following

6.11 PROPOSITION. Assume that $m = |G:U|$ is finite. Fix a transversal $\{x_k\}$ as in (6.1) and let $\bar{g}$ denote the representative of the coset Ug. Assume that A is a left and B a right G-module. Then $Cor_n\colon H_n(G,B) \to H_n(U, VB)$ is induced by the chain transformation

$$\{ \varphi_n : B \otimes_G B_n(G) \to VB \otimes_U B_n(U) \},$$

on generators given by

$$\varphi_n(b \otimes (g_o, \ldots, g_n)) = \sum_{k=1}^{m} b\cdot x_k^{-1} \otimes (x_k g_o \cdot \overline{x_k g_o}^{-1}, \ldots, x_k g_n \cdot \overline{x_k g_n}^{-1}),$$

while $Cor^n\colon H^n(U,VA) \to H^n(G,A)$ is induced by the cochain transformation

$$\{ \varphi^n : Hom_U(B_n(U), VA) \to Hom_G(B_n(G), A) \}$$

with

$$\varphi^n(f)(g_o, \ldots, g_n) = \sum_{k=1}^{m} x_k^{-1}\cdot f(x_k g_o \cdot \overline{x_k g_o}^{-1}, \ldots, x_k g_n \cdot \overline{x_k g_n}^{-1}). \quad \square$$

6.12 COROLLARY. (a) With $\rho\colon H_1G \cong G_{ab}$ as in Proposition 5.8, $\rho\circ Cor_1\circ\rho^{-1}\colon G_{ab} \to U_{ab}$ is given by

$$| \; g[G,G] \longmapsto \prod_{k=1}^{m} x_k g \cdot \overline{x_k g}^{-1} \cdot [U,U] \; |$$

(b) $Cor^2: H^2(U,VA) \to H^2(G,A)$ maps the class of the factor set $f: U \times U \to VA$ onto the class of the factor set

$$\varphi^2(f)[x|y] = \sum_{k=1}^{m} x_k^{-1} \cdot f[x_k x \cdot \overline{x_k x}^{-1} | \overline{x_k x} \cdot y \cdot \overline{x_k xy}^{-1}] \; .$$

The formula of (a) describes the <u>transfer</u> homomorphism (<u>Verlagerung</u>) τ of Burnside and Schur, cf. ZASSENHAUS [1; p.167] and HUPPERT [1; IV.1.4 (b)].

PROOF. This amounts to rewriting the formulas of Proposition 6.11 for $n = 1$ and $n = 2$ in terms of the inhomogeneous generators $[g] = (1,g)$ and $[x|y] = (1,x,xy)$ of the bar resolution. Note that the $[x]$ in Proposition 5.8 here reads as $1 \otimes [x] \in Z \otimes_Q B_1(Q)$. □

CHAPTER II. SCHUR'S THEORY OF PROJECTIVE REPRESENTATIONS

1. Projective Representations

Throughout this section let K be a field, K^* its multiplicative group, V a vector space over K of arbitrary dimension, while $GL(V)$ denotes the group of linear automorphisms of V.

1.1 DEFINITION. A projective representation of a group Q over the vector space V is a map

$$P : Q \to GL(V)$$

which satisfies

$$P(g)\, P(h) = P(gh)\, \omega(g,h), \quad \omega(g,h) \in K^*$$

for all $g,h \in Q$. The map $\omega: Q\times Q \to K^*$ is called the corresponding factor system.

Quite often projective representations appear in connection with irreducible linear representations. Assume for the moment that K is algebraically closed, and let $\beta: G \to GL(V)$ be an irreducible linear representation of a group G, where V is finite dimensional. If N is a central subgroup of G, then the elements of N are mapped onto dilatations by Schur's lemma. Let T be a transversal to N in G. Then each element $g \in G$ has a unique decomposition $g = nt$, $n \in N$, $t \in T$, and the map $P: G/N \to GL(V)$ defined by $P(tN) := \beta(t)$ is a projective representation. A more general concept which describes the connection between representations of groups and normal subgroups of finite index is contained in CLIFFORD [1], see also CURTIS/REINER [1; §49, §51], and HUPPERT [1; V §17].

Now we return to the projective representation given in Definition 1.1. The associativity of the multiplication in GL(V) easily yields the relation I.(1.6) for ω, where K^* is regarded as a Q-module with trivial action. The factor system is normalized in the sense of I.1.1, if and only if P maps the identity element of Q onto the identity of V.

1.2 DEFINITION. Let V_1, V_2 be vector spaces over K, and P_1, P_2 projective representations of Q over V_1, resp. V_2 with corresponding factor systems ω_1, ω_2.

(i) P_1 and P_2 are called <u>(projectively) equivalent</u>, if there exists an isomorphism $\beta: V_1 \to V_2$ and a map $c: Q \to K^*$ which satisfy

$$\beta\, P_1(g) = c(g)\, P_2(g)\, \beta \ , \ g \in Q \ .$$

(ii) If the preceding condition can be satisfied with $c(g) = 1$ for all $g \in Q$, then P_1 and P_2 are called <u>linearly equivalent</u>.

Assume that P_1 and P_2 are equivalent. Then an easy calculation shows

$$(1.1) \quad \omega_1(g,h)\, c(gh) = \omega_2(g,h)\, c(g)\, c(h) \ , \ g,h \in Q \ ,$$

i.e. the corresponding factor systems differ by a principal one, cf. I.1.1, and we call them <u>equivalent</u> as well. Linearly equivalent projective representations have the same factor system.

On the other hand, if ω_1 and ω_2 are equivalent factor systems satisfying (1.1), then

$$P \longmapsto cP$$

is a one-to-one correspondence between the projective representations of Q with factor system ω_1 and those with factor system ω_2, respecting projective and linear equivalence.

Assume that $c, c': Q \to K^*$ are two maps satisfying (1.1). Then $c' \cdot c^{-1}$ is a homomorphism, in other words a one-dimensional character of Q. Now we fix two equivalent factor systems ω_1 and ω_2 and a map c, leading to the above correspondence for projective representations. Assume that P_i is a projective representation with factor system ω_i , $i = 1,2$, and let P_1 and P_2 be equivalent. Then by 1.1 there exists a map $c': Q \to K^*$ such that $c'P_1$ and P_2 are linearly equivalent. Furthermore we have

$$c'P_1 = (c'c^{-1})(cP_1) \quad \text{and} \quad (c'c^{-1}) \in \mathrm{Hom}(Q,K^*) \ .$$

Hence, the classes under projective equivalence of those projective representations, whose factor systems belong to a fixed equivalence class represented by a factor system ω, are given by the orbits of the character group Hom(Q,K*) acting by multiplication on the classes with respect to linear equivalence of projective representations with factor system ω.

1.3 REMARK. Beside Definition 1.1 there are two other ways describing projective representations.

(i) One of them uses the notion of the twisted group algebra. For each factor system ω of Q over K we can consider the set of all finite formal sums $\sum a_g g$, $a_g \in K$, $g \in Q$ endowed with the multiplication $\circ$ generated by

$$g \circ h = \omega(g,h)gh \ .$$

This yields an associative K-algebra $(KQ)_\omega$ of dimension $|Q|$, called the <u>twisted group algebra</u>, and it easily follows that there is a one-to-one correspondence between its modules and the projective representations of Q with factor system ω. The isomorphism of modules corresponds to the linear equivalence of projective representations.

(ii) For each K-vector space V we consider the following central extension:

$$(1.2) \qquad \sigma_V : K^* \overset{\delta}{\rightarrowtail} GL(V) \overset{\tau}{\twoheadrightarrow} PGL(V) ,$$

where $\delta(k)$ is the dilatation $v \mapsto kv$, for all $k \in K^*$. The image of δ is the center of GL(V) , and the projective linear group PGL(V) is defined as the factor group with canonical projection τ.

Let $P: Q \to GL(V)$ be a projective representation. Then $\gamma = \tau P: Q \to PGL(V)$ is a homomorphism. On the other hand, if $\gamma: Q \to PGL(V)$ is a homomorphism, any map $P: Q \to GL(V)$ satisfying $\tau P = \gamma$ is a projective representation of Q over V. Let $\gamma_i: Q \to PGL(V_i)$, i=1,2 be homomorphisms. If there exists a K-isomorphism $\beta: V_1 \to V_2$ such that the induced isomorphism $\beta_*: PGL(V_1) \to PGL(V_2)$ satisfies $\beta_* \gamma_1 = \gamma_2 \beta_*$, then we call γ_1 and γ_2 equivalent. It is easy to see that projective representations P_1, P_2 with $\tau_1 P_1 = \gamma_1$, $\tau_2 P_2 = \gamma_2$ are projectively equivalent, if and only if γ_1 and γ_2 are equivalent. Hence, if we are solely interested in projective representations up to projective equivalence, we can regard them as homomorphisms into projective groups. In view of the preceding, we usually do not strictly distinguish between projective representations (in the sense of Definition 1.1) and homomorphisms into projective groups, which will be called projective representations as well.

1.4 DEFINITION. Let $\gamma: Q \to PGL(V)$ be a projective representation of Q. Then $\gamma^*(\sigma_V) \in \mathrm{Cext}(Q,K^*)$ is called the <u>cohomology element</u> associated with γ.

Let us recall that $\gamma^*(\sigma_V)$ is represented by the backward induced extension $\sigma_V \gamma: K^* \rightarrowtail G \twoheadrightarrow Q$, which yields a morphism $(1,\beta,\gamma): \sigma_V \gamma \to \sigma_V$ of extensions as in (1.3) and is determined by this property up to congruence, cf. Section I.1:

$$
(1.3)\qquad
\begin{array}{rccccc}
\sigma_V\gamma : & K^* & \overset{\lambda}{\rightarrowtail} & G & \overset{\rho}{\twoheadrightarrow} & Q \\
& \Vert & & \downarrow{\scriptstyle\beta} & & \downarrow{\scriptstyle\gamma} \\
\sigma_V : & K^* & \overset{\delta}{\rightarrowtail} & GL(V) & \overset{\tau}{\twoheadrightarrow} & PGL(V)
\end{array}
$$

1.5 PROPOSITION. Let $\gamma: Q \to PGL(V)$ be a homomorphism (projective representation), and $P: Q \to GL(V)$ a map with $\tau P = \gamma$, and let ω denote the factor system of P. Then the element of $H(Q,K^*,0)$ represented by ω corresponds to $\gamma^*(\sigma_V)$, cf. I.1.1.

PROOF. Let G denote the Cartesian product of K^* and Q with the multiplication

$$(k_1,q_1)(k_2,q_2) = (k_1k_2\omega(q_1,q_2),q_1q_2) .$$

Then we have the central extension

$$e : K^* \overset{\varkappa}{\rightarrowtail} G \overset{\pi}{\twoheadrightarrow} Q ,$$

where $\varkappa = (k \longmapsto (k\omega(1,1),1))$, $\pi = ((k,q) \longmapsto q)$, i.e. e is an extension corresponding to ω. We also have a homomorphism $\alpha: G \to GL(V)$ defined by

$$\alpha(k,q) = \delta(k) \cdot P(q) ,$$

such that the following diagram is commutative:

$$
(1.4)\qquad
\begin{array}{rccccc}
e : & K^* & \overset{\varkappa}{\rightarrowtail} & G & \overset{\pi}{\twoheadrightarrow} & Q \\
& \Vert & & \downarrow{\scriptstyle\alpha} & & \downarrow{\scriptstyle\gamma} \\
\sigma_V : & K^* & \overset{\delta}{\rightarrowtail} & GL(v) & \overset{\tau}{\twoheadrightarrow} & PGL(V) ,
\end{array}
$$

which has the form (1.3). □

1.6 PROPOSITION. There exists a vector space V of dimension $|Q|$ such that the map $\gamma \longmapsto \gamma^*(\sigma_V): Hom(Q,PGL(V)) \to Cext(Q,K^*)$ is surjective.

PROOF. There are two obvious ways of proving 1.6. Let $x \in \mathrm{Cext}(Q,K^*)$, and ω a factor system representing x. Then we can form the twisted group algebra $(KQ)_\omega$, and consider its regular left module. This module gives rise to a projective representation of Q with factor system ω and a vector space of dimension $|Q|$, and the assertion follows from 1.5.

Alternatively let

$$e : K^* \overset{\varkappa}{\rightarrowtail} G \overset{\pi}{\twoheadrightarrow} Q$$

be a central extension with $[e] = x \in \mathrm{Cext}(Q,K^*)$. Then $\varkappa^{-1}$ is a one-dimensional, linear representation of $U = \varkappa(K^*)$, and we denote by α the representation of G which is induced by $\varkappa^{-1}$ and has representation space $KG \otimes_U K$ of dimension $|Q|$, cf. CURTIS/REINER [1;p.73]. As e is central, α maps the elements of U onto dilatations, which yields a commutative diagram (1.4), and proves $x = [e] = \gamma^*(\sigma_V)$. In fact, both constructions above yield equivalent projective representations. □

1.7 REMARK. A projective representation $P: Q \to GL(V)$, resp. $\gamma = \tau P$ is called <u>irreducible</u>, if 0 and V are the only subspaces of V that are invariant under all $P(g)$, $g \in Q$. As the regular left module of any ring with unit element contains maximal submodules, the first proof of 1.6 shows that each element of Cext(Q,K*) can be realized by an irreducible projective representation. In Section 3 we shall see that finite dimensional projective representations yield only elements of finite order in the abelian group Cext(Q,K*). This is the reason why we consider representations of arbitrary dimension.

Let us return to a situation considered at the beginning of this section. If $\beta : G \to GL(V)$ is a finite dimensional irreducible representation of a group over an algebraically closed field K and $e : N \rightarrowtail G \twoheadrightarrow Q$ a central extension, we have obtained a projective representation P of Q by Schur's Lemma. Then $\gamma = \tau P: Q \to PGL(V)$ fits in the commutation diagram (1.5)

$$\begin{array}{ccccccc} e: & N & \rightarrowtail & G & \twoheadrightarrow & Q \\ & \downarrow{\scriptstyle\alpha} & & \downarrow{\scriptstyle\beta} & & \downarrow{\scriptstyle\gamma} \\ \sigma_V: & K^* & \overset{\delta}{\rightarrowtail} & GL(V) & \twoheadrightarrow & PGL(V) \end{array} \qquad (1.5)$$

In the following we say, whenever a commutative diagram of the form (1.5) is given, that the projective representation γ can be <u>lifted</u> in e. (Here e need not be central.) As σ_V is central for all V, γ can be lifted in e, if and only if it can be lifted in c(e), the centralisation of e. Thus we restrict our attention in the following to central extensions in order to lift projective representations. From 1.5, resp. I.1.12 we obtain that $\gamma: Q \to PGL(V)$ can be lifted in Q itself, i.e. in the trivial extension $0 \rightarrowtail Q \twoheadrightarrow Q$, if and only if its cohomology element $\gamma^*(\sigma_V)$ vanishes, i.e. if the corresponding projective representations $P: Q \to GL(V)$ with $\tau P = \gamma$ are equivalent to linear representations. In general we obtain:

1.8 PROPOSITION. A projective representation $\gamma: Q \to PGL(V)$ can be lifted in the central extension $e: N \rightarrowtail G \twoheadrightarrow Q$, if and only if there is a homomorphism $\alpha: N \to K^*$ with $\gamma^*(\sigma_V) = \theta^*(e,K^*)\alpha$.

PROOF. Recall from I.2.7 that $\theta^*(e,K^*): Hom(N,K^*) \to Cext(Q,K^*)$ is the map $\alpha \longmapsto [\alpha e]$. Now the proof of the proposition is immediate from Theorem I.1.10. □

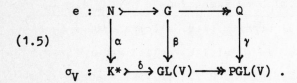

Propositions 1.6 and 1.8 clearly show that the question whether a certain projective representation can be lifted in a given central extension e, is not a question of representation theory, but depends only on the image of $\theta^*(e,K^*)$, hence on the structure of e and the

abelian group K^*. So we may replace σ_V in (1.5) by any other central extension with kernel K^*. It is only important (by I.1.10) that the extension which is backward induced by the homomorphism on the right side lies in the image of $\theta^*(e,K^*)$. The study of lifting homomorphisms will be continued in Section 2. Extensions e having the property that $\theta^*(e,K^*)$ is an epimorphism (i.e. all projective K-representations of Q can be lifted in e) will be called generalized K^*-representation groups, cf. Definition 2.2. Trivial examples are given by the centralizations

$$R/[F,R] \rightarrowtail F/[F,R] \twoheadrightarrow Q$$

of free presentations $R \rightarrowtail F \twoheadrightarrow Q$.

Now we consider a fixed central extension

$$e : N \overset{\varkappa}{\rightarrowtail} G \overset{\pi}{\twoheadrightarrow} Q ,$$

and describe the projective representations of Q which can be lifted in e, in terms of linear representations of G. Let $T = \{u_q \mid q \in Q\}$ be a transversal to $\varkappa N$ in G, and $f: Q\times Q \to N$ be the associated factor system, cf. I.1.1. For each $\alpha \in \mathrm{Hom}(N,K^*)$ we define the factor system

$$\omega(\alpha,T) = \alpha f : Q\times Q \to K^* ,$$

and it readily follows that $\omega(\alpha,T)$ corresponds to $\theta^*(e,K^*)\alpha$, cf. I.1.5. Hence by 1.8 each projective representation that can be lifted in e, has a factor system equivalent to some $\omega(\alpha,T)$.

1.9 PROPOSITION. Let $\alpha_i \in \mathrm{Hom}(N,K^*)$,i=1,2. (i) If e is a stem extension, i.e. $\varkappa N \subseteq [G,G]$, $\omega(\alpha_1,T)$ and $\omega(\alpha_2,T)$ are equivalent, if and only if $\alpha_1 = \alpha_2$.

(ii) If K is algebraically closed, $\omega(\alpha_1,T)$ and $\omega(\alpha_2,T)$ are equivalent, if and only if $\alpha_1\varkappa^{-1}$ and $\alpha_2\varkappa^{-1}$ coincide on $\varkappa N \cap [G,G] = \varkappa(\mathrm{Im}\ \theta_*(e))$.

PROOF. Theorem I.2.7 yields that any two factor systems $\omega(\alpha_1,T)$ and $\omega(\alpha_2,T)$ are equivalent, if and only if $(\alpha_1-\alpha_2)\varkappa^{-1} = (\varkappa n \longmapsto \alpha_1(n)\ \alpha_2(n)^{-1}) : \varkappa N \to K^*$ can be extended to G. Now the proofs of (i) and (ii) follow easily. For the second statement one has to keep in mind that K* is a divisible abelian group, if K is algebraically closed. □

Let $\beta: G \to GL(V)$ be a linear K-representation of G which maps the elements of $\varkappa N$ onto dilatations. Hence we have

(1.6) $\beta(\varkappa n) = v \longmapsto \alpha(n)v$

for some $\alpha \in Hom(N,K^*)$, and we obtain a projective representation of Q by

$$P(\beta,T) = q \longmapsto \beta(uq)$$

having factor system $\omega(\alpha,T)$.

1.10 PROPOSITION. (i) A projective K-representation can be lifted in e, if and only if it is projectively equivalent to some $P(\beta,T)$.

(ii) If β runs through all linear representations of G satisfying (1.6) for a fixed α, then $P(\beta,T)$ runs through all projective representations of Q with factor system $\omega(\alpha,T)$.

(iii) Let β_1 and β_2 satisfy (1.6) with the same $\alpha \in Hom(N,K^*)$. Then β_1 and β_2 are equivalent, if and only if $P(\beta_1,T)$ and $P(\beta_2,T)$ are linearly equivalent.

(iv) Let β_1 and β_2 be as in (iii). Then $P(\beta_1,T)$ and $P(\beta_2,T)$ are projectively equivalent, if and only if there exists $\chi \in Hom(Q,K^*)$ such that $(\chi\pi)\beta_2 = \{g \longmapsto (\chi\pi g)\beta_1(g)\}$ and β_2 are equivalent (as linear K-representations).

PROOF. Assertion (i) is just a slight reformulation of the definition of lifting projective representations, cf. (1.5), and (iii) is trivial. As mentioned above, $\omega(\alpha,T)$ is the factor system of each $P(\beta,T)$. On the other hand let P be any projective representation with factor system $\omega(\alpha,T)$. Let $\gamma = \tau P$ be the corresponding homomorphism into the projective group. Then 1.9 gives rise to a linear representation β of G with commutative diagram (1.5). Thus we obtain $P(q) = \chi(q)\, P(\beta,T)$ with a map $\chi\colon Q \to K^*$. As P and $P(\beta,T)$ have the same factor system $\omega(\alpha,T)$, we have $\chi \in \mathrm{Hom}(Q,K^*)$. This proves $P = P((\chi\pi)\beta,T)$ and completes the proof of (ii) and (iv).

□

2. The Problem of Lifting Homomorphisms

In this section, we fix an abelian group A and a central extension $e = (\varkappa,\pi): N \rightarrowtail G \twoheadrightarrow Q$. We study the problem whether there exist commutative diagrams

$$
(2.1)\qquad
\begin{array}{cccccc}
e: & N & \rightarrowtail & G & \twoheadrightarrow & Q \\
 & \downarrow{\scriptstyle\alpha} & & \downarrow{\scriptstyle\beta} & & \downarrow{\scriptstyle\gamma} \\
e': & A & \rightarrowtail & G' & \twoheadrightarrow & Q'
\end{array}
$$

for a given homomorphism γ and all central extensions e' in some specified class Φ. In other words, can γ be lifted to a morphism $(\cdot,\cdot,\gamma): e \to e'$ of extensions for all $e' \in \mathrm{Cext}(Q',A) \cap \Phi$?
In the previous section we considered the case $A = K^*$ and $\Phi = \{\sigma_V\}$, thus γ being a projective representation over the fixed K-vector space V. We now say that $\gamma: Q \to Q'$ can be <u>lifted</u> in e <u>completely</u> (with respect to A), if the problem has a solution for all extensions of A, i.e. Φ contains Cext(Q',A).

According to Theorems I.1.10 and I.2.7 (combined as in the proof of Proposition 1.8), our problem has a positive answer, if and only if $\gamma^*(e') \in \mathrm{Im}\ \theta^*(e,A)$ for all $e' \in \mathrm{Cext}(Q',A) \cap \Phi$. Thus it is the subgroup $\mathrm{Im}\ \theta^*(e,A)$ of Cext(Q,A) which controls "the lifting properties" of e. In particular, we have:

2.1 PROPOSITION. Every homomorphism $\gamma: Q \to Q'$ can be lifted completely with respect to A (now Q' varying), precisely when $\theta^*(e,A): \mathrm{Hom}(N,A) \to \mathrm{Cext}(Q,A)$ is an epimorphism. □

2.2 DEFINITION, cf. YAMAZAKI [1;§3.2]. (i) The central extension e is a <u>generalized</u> <u>A-representation</u> <u>group</u> resp. an A-<u>representation</u>

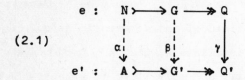

group of Q, if $\theta^*(e,A)$ is an epimorphism resp. isomorphism.

(ii) In case $A = \mathbb{C}^*$, where $\mathbb{C}$ denotes the complex numbers, a (generalized) $\mathbb{C}^*$-representation group is simply called a (generalized) representation group of Q.

2.3 REMARK. We'll see below that $\mathbb{C}$ in 2.2(ii) could be replaced by an arbitrary algebraically closed field of characteristic zero. It follows from Proposition 1.6, for any field K, that all projective K-representations of Q can be lifted if, and only if, e is a generalized K^*-representation group of Q. This answers our initial problem for $A = K^*$ and $\Phi = \{\sigma_V \mid \text{all } K\text{-vector spaces } V\}$.

In Section 1 we gave (trivial) examples of extensions which are generalized A-representation groups of Q for all A. However, A-representation groups of Q do not always exist. For example, let $Q = \mathbb{Z}/2$ and $A = \mathbb{Z}$; then

$$|\mathrm{Cext}(Q,A)| = 2 \neq |\mathrm{Hom}(N,A)|$$

whatever abelian group N is chosen. The concept of generalized A-representation groups behaves nicely with respect to epimorphisms:

2.4 PROPOSITION. Assume that e is a generalized A-representation group of Q. Let $\rho\colon Q \to \bar{Q}$ be an epimorphism, and $K = \pi^{-1}(\mathrm{Ker}\ \rho)$. Then $\rho\pi$ induces a generalized A-representation group

$$\mathrm{Ker}(\rho\pi)' \hookrightarrow \frac{G}{[G,K]} \xrightarrow{(\rho\pi)'} \bar{Q}$$

of $\bar{Q}$.

PROOF. Let $\bar{e}$ denote the extension above, which is central by definition. Let $e'=(\cdot,\sigma)\colon A \rightarrowtail G' \twoheadrightarrow Q'$ be an arbitrary central extension and $\gamma\colon \bar{Q} \to Q'$ a homomorphism. The extension e is a generalized A-representation group, and by 2.1 and 2.2 we have a morphism $(\cdot,\beta',\gamma\rho)\colon e \to e'$. Thus $\sigma\beta' = \gamma\rho\pi$, which implies $\beta'(K) \subseteq \operatorname{Ker}\sigma$. As $\operatorname{Ker}\sigma$ is central in G', we obtain $[G,K] \subseteq \operatorname{Ker}\beta'$, and β' induces a morphism $(\cdot,\beta,\gamma)\colon \bar{e} \to e'$. Hence, each homomorphism of $\bar{Q}$ can be lifted in $\bar{e}$ completely; i.e. $\bar{e}$ is a generalized A-representation group. □

Our fixed data e,A give rise to the following diagram

(2.2)
$$\begin{array}{ccccc} & & \operatorname{Hom}(N,A) & & \\ & & \downarrow{\scriptstyle \theta^*(e,A)} & \searrow^{\theta_*(e)^*} & \\ \operatorname{Ext}(Q_{ab},A) & \overset{\Psi}{\rightarrowtail} & \operatorname{Cext}(Q,A) & \overset{\theta_*}{\twoheadrightarrow} & \operatorname{Hom}(M(Q),A) \end{array},$$

the exact row of which is I.(3.7) of the Universal Coefficient Theorem, while $\theta_*(e)^*$ is defined as the indicated composite map. By Lemma I.3.7, we have $\theta_*(\theta^*(e,A)\alpha) = \alpha\theta_*(e)$ for all $\alpha \in \operatorname{Hom}(N,A)$. Thus

$$\theta_*(e)^* = \operatorname{Hom}(\theta_*(e),A) = (\alpha \longmapsto \alpha\theta_*(e))\colon \operatorname{Hom}(N,A) \to \operatorname{Hom}(M(Q),A),$$

the dependence on A being suppressed from the notation.

2.5 PROPOSITION. (i) If e is a generalized A-representation group, then $\theta_*(e)^*$ is an epimorphism.

(ii) Let e_1 and e_2 be central extensions with factor group Q. If in e_1 and e_2 the same homomorphisms of Q are liftable with respect to the same central extensions of A, then $\operatorname{Im}\theta_*(e_1)^* = \operatorname{Im}\theta_*(e_2)^*$.

PROOF. The homomorphism $\theta^*(e,A)$ is an epimorphism, if e is a generalized representation group, and (i) follows from diagram (2.2). The arguments leading to 1.8 and 2.1 show that e_1 and e_2 have the same lifting properties if, and only if $\operatorname{Im}\theta^*(e_1,A) = \operatorname{Im}\theta^*(e_2,A)$. Hence, (ii) follows from (2.2). □

In the following we quite often study the important special case where $\mathrm{Ext}(Q_{ab},A) = 0$. From the considerations above we easily obtain

2.6 PROPOSITION. (i) If $\mathrm{Ext}(Q_{ab},A) = 0$, the converse in 2.5(i) and 2.5(ii) holds.

(ii) The following statements are equivalent:

(a) e is a generalized A-representation group and $\theta_*(e)^*$ an isomorphism.

(b) e is an A-representation group and $\mathrm{Ext}(Q_{ab},A) = 0$

(c) $\theta_*(e)^*$ is an isomorphism and $\mathrm{Ext}(Q_{ab},A) = 0$. □

2.7 LEMMA. Let $L(Q,A)$ denote the intersection of the subgroups $\mathrm{Ker}\,\alpha \subseteq M(Q)$, where α runs through $\mathrm{Hom}(M(Q),A)$. Then $\theta_*(e)^*$ is an epimorphism, if, and only if the following two conditions hold:

(a) $L(Q,A) \supseteq \mathrm{Ker}\,\theta_*(e)$

(b) Each homomorphism $\alpha': \mathrm{Im}\,\theta_*(e) \to A$ can be extended to N.

Furthermore, $\theta_*(e)^*$ is an isomorphism, if, and only if we have in addition:

(c) $\mathrm{Hom}(\frac{\varkappa N}{\varkappa N \cap [G,G]},A) = 0$.

Condition (b) is satisfied if $\mathrm{Ext}(\frac{\varkappa N}{\varkappa N \cap [G,G]},A) = 0$, and (b) and (c) hold if e is a stem extension, cf. I.3.11.

PROOF. The homomorphism $\theta_*(e)$ is epimorphic, if and only if for each $\alpha \in \mathrm{Hom}(M(Q),A)$ there exists $\beta \in \mathrm{Hom}(N,A)$, such that $\alpha = \beta\theta_*(e)$. Assume (a) and (b). By (a), each $\alpha \in \mathrm{Hom}(M(Q),A)$ induces a homomorphism $\alpha': \mathrm{Im}\,\theta_*(e) \to A$ with $\alpha'\theta_*(e) = \alpha$. By (b) we have a $\beta \in \mathrm{Hom}(N,A)$ extending α', and which therefore satisfies $\alpha = \beta\theta_*(e)$. On the other hand, let $\theta_*(e)^*$ be epimorphic. Then the relation $\alpha = \beta\theta_*(e)$ for all $\alpha \in \mathrm{Hom}(M(Q),A)$ yields

$L(Q,A) \supseteq \operatorname{Ker} \theta_*(e)$. Let $\alpha' \in \operatorname{Hom}(\operatorname{Im} \theta_*,A)$. Then we have a homomorphism $\beta: N \to A$ with $\alpha'\theta_*(e) = \beta\theta_*(e)$. Thus β extends α' to N, proving the first part of the Lemma. The homomorphism $\theta_*(e)^*$ is injective, if and only if $\operatorname{Hom}(\operatorname{Coker} \theta_*(e),A) = 0$. As $\operatorname{Im} \theta_*(e) = \varkappa^{-1}(\varkappa N \cap [G,G])$, we have

$$\operatorname{Coker} \theta_*(e) \cong \frac{\varkappa N}{\varkappa N \cap [G,G]}$$

which proves the second assertion. The remaining part of the proof is trivial. □

2.8 COROLLARY. If A is a divisible abelian group, then

(i) e is a generalized A-representation group, if and only if $\operatorname{Hom}(\operatorname{Ker} \theta_*(e),A) = 0$.

(ii) e is an A-representation group, if and only if $\operatorname{Hom}(\operatorname{Ker} \theta_*(e),A) = 0$ and $\operatorname{Hom}(\operatorname{Coker} \theta_*(e),A) = 0$.

PROOF. Assume that A is divisible. Hence $\operatorname{Ext}(Q_{ab},A) = 0$, and we can apply 2.5 and 2.6; i.e. e is a generalized A-representation group, resp. A-representation group, if and only if $\theta_*(e)^*$ is an epimorphism, resp. an isomorphism. As A is divisible, condition (b) in 2.7 is satisfied, whereas (a) is equivalent to $\operatorname{Hom}(\operatorname{Ker} \theta_*(e),A) = 0$. The condition $\operatorname{Hom}(\operatorname{Coker} \theta_*(e),A) = 0$ is a reformulation of (c) in 2.7, and we are done. □

From 2.5, 2.6 and 2.7 we obtain that a generalized A-representation group, which is a stem extension, is even an A-representation group. This can only hold if $\operatorname{Ext}(Q_{ab},A) = 0$, but in this case I.3.8 guaranties the existence of such extensions, e.g. choose e with $N = M(Q)$ and $\theta_*(e) = 1_{M(Q)}$.

Let us return to the general case, where A is an arbitrary

abelian group, and assume that e is a stem extension. Then $\theta_*(e)^*$ is a monomorphism and we obtain from (2.2)

$$\mathrm{Im}\ \theta^*(e,A) \cap \Psi\ \mathrm{Ext}(Q_{ab},A) = 0 .$$

This relation and Lemma 2.7 yield

2.9 PROPOSITION. Let e be a stem extension. Then the following properties are equivalent:

(i) $L(Q,A) \supseteq \mathrm{Ker}\ \theta_*(e)$

(ii) $\theta_*(e)^*$ is an isomorphism

(iii) Cext(Q,A) is the internal direct sum of $\mathrm{Im}\ \theta^*(e,A)$ and $\Psi\ \mathrm{Ext}(Q_{ab},A)$. □

2.10 EXAMPLE. The preceding proposition can be applied as follows. For every group Q and every abelian group A, we "construct" generalized A-representation groups of Q in a different manner than before. The abelian groups Hom(N,A) and Cext(Q,A) are End(A)-modules by $\alpha \longmapsto \varphi\circ\alpha$ and $[e'] \longmapsto \varphi_*[e'] = [\varphi e']$, $\alpha \in \mathrm{Hom}(N,A)$, $[e'] \in \mathrm{Cext}(Q,A)$, $\varphi \in \mathrm{End}(A)$. The calculus of induced extensions implies that $\Psi\ \mathrm{Ext}(Q_{ab},A)$ is a submodule of Cext(Q,A) and $\theta^*(e,A)$ is an End(A)-homomorphism. Let $[e_i]_{i\in I}$ be a system of End(A)-generators of $\Psi\ \mathrm{Ext}(Q_{ab},A)$

$$e_i : A \stackrel{\varkappa_i}{\rightarrowtail} G_i \stackrel{\pi_i}{\twoheadrightarrow} Q .$$

Let $\prod_{i\in I} G_i$ be the (unrestricted) direct product of the groups G_i, and let

$$G^1 = \{x \mid x=(x_i)_{i\in I} \in \prod_{i\in I} G_i ,\ \pi_i(x_i) = \pi_\gamma(x_\gamma) \text{ for all } i,j \in I\}$$

be their fibre product. Hence we obtain a central extension

$$e^1 : \prod_{i\in I} A \rightarrowtail G^1 \twoheadrightarrow Q .$$

For two extensions e_1, e_2 we obtain $(e_1\times e_2)\Delta_Q$, which is part of the sum defined in Cext(Q,A), cf. I.(2.2). Let $\rho_j : \prod_{i \in I} A \twoheadrightarrow A$ be the j-th projection. Then we have $\rho_j e^1 = e_j$. Hence, $\theta^*(e^1,A)$ maps $\mathrm{Hom}(\Pi A,A)$ onto $\Psi\,\mathrm{Ext}(Q_{ab},A)$. Let $e^2 : N^2 \rightarrowtail G^2 \twoheadrightarrow Q$ be a stem extension satisfying one, and hence all the conditions of 2.9, and consider

$$e_o = (e^1 \times e^2)\Delta_Q : \prod_{i \in I} A \times N^2 \rightarrowtail G^1 \curlywedge G^2 \twoheadrightarrow Q .$$

Then we have

$$\mathrm{Hom}(\Pi A \times N^2,A) = \mathrm{Hom}(\Pi A,A) \times \mathrm{Hom}(N^2,A)$$

and

$$\theta^*(e_o,A) = \theta^*(e^1,A) \times \theta^*(e^2,A) .$$

As $\theta^*(e^1,A)$ maps onto $\Psi\,\mathrm{Ext}(Q_{ab},A)$ and $\mathrm{Im}\,\theta^*(e^2,A)$ is assumed to be a complement of $\Psi\,\mathrm{Ext}(Q_{ab},A)$ in Cext(Q,A), $\theta^*(e_o,A)$ is an epimorphism, and thus e_o is a generalized A-representation group.

2.11 EXAMPLE. Assume $Q = S_n$, the symmetric group on n letters, $n \geq 3$ and $A = \mathbb{Z}/2$, which can be regarded as the multiplicative group of GF(3). We have $(S_n)_{ab} \cong M(S_n) \cong \mathbb{Z}/2$, and we can identify $A = M(Q) = \mathbb{Z}/2$, cf. 3.8. Thus we obtain

$$\mathrm{Ext}(Q_{ab},A) \cong \mathbb{Z}/2 , \ \mathrm{Hom}(M(Q),A) \cong \mathbb{Z}/2 , \ \mathrm{Cext}(Q,A) \cong \mathbb{Z}/2 \times \mathbb{Z}/2 .$$

A generator of $\Psi\,\mathrm{Ext}(Q_{ab},A)$ is given by

$$e^1 : \mathbb{Z}/2 \rightarrowtail S_n \curlywedge \mathbb{Z}/4 \twoheadrightarrow S_n ,$$

where $(S_n)_{ab}$ and $(\mathbb{Z}/4)/2(\mathbb{Z}/4)$ are identified. Choose $e^2 : \mathbb{Z}/2 \rightarrowtail D \twoheadrightarrow S_n$ such that $\theta_*(e_2) = 1_{\mathbb{Z}/2}$. From the consideration in 2.10 we obtain a central extension

$$e_o : \mathbb{Z}/2 \times \mathbb{Z}/2 \rightarrowtail D \curlywedge (S_n \curlywedge \mathbb{Z}/4) \twoheadrightarrow S_n ,$$

which is a generalized A-representation group for S_n. As the image and the domain of $\theta^*(e_o,A)$ have the same order, $\theta^*(e_o,A)$ is an isomorphism, i.e. e_o is even an A-representation group, and it is a minimal extension in which all projective GF(3)-representations of

S_n can be lifted. It is easy to see that the representation which maps a generator of $\mathbb{Z}/4$ onto the matrix $\begin{pmatrix} 0 & -1 \\ 1 & 0 \end{pmatrix}$ over GF(3) induces an irreducible projective representation of S_n, whose cohomology element is a generator of $\Psi\,\mathrm{Ext}(Q_{ab},A)$, and which therefore cannot be lifted in any stem extension of S_n.

In the following we restrict our attention to the case where $\mathrm{Ext}(Q_{ab},A) = 0$. Then the (generalized) A-representation groups are characterized by 2.5, 2.6 and 2.7. As mentioned above, A-representation groups for all A with $\mathrm{Ext}(Q_{ab},A) = 0$ are given by the special stem extensions e satisfying $N = M(Q)$ and $\theta_*(e) = 1_{M(Q)}$. Let e be any A-representation group for Q, and B an abelian group with $\mathrm{Hom}(B,A) = 0$. Then

$$e' : N \times B \rightarrowtail G \times B \twoheadrightarrow Q$$

is also an A-representation group. Hence, an A-representation group (in the sense of Definition 2.2) need not be a stem extension, not even if $\mathrm{Ext}(Q_{ab},A) = 0$.

2.12 REMARK. Assume $\mathrm{Ext}(Q_{ab},A) = 0$ and $\mathrm{Hom}(N/U,A) \neq 0$ for all proper subgroups U of N. Then e is an A-representation group for Q, if and only if e is a stem extension satisfying $L(Q,A) \supseteq \mathrm{Ker}\,\theta_*(e)$.

PROOF. It remains to show that each A-representation group e has to be a stem extension. If $\theta_*(e)$ were not surjective, there are at least two elements of $\mathrm{Hom}(N,A)$ vanishing on $\mathrm{Im}\,\theta_*(e)$. Hence, $\theta_*(e)^*$ cannot be injective, a contradiction. □

The group theoretical significance of the condition $\mathrm{Ext}(Q_{ab},A)= 0$ was given in I.3.13. As mentioned above, it is satisfied for all Q, if A is a divisible abelian group, and in this case the A-representation groups were described in 2.8. Examples for divisible groups are the multiplicative groups of algebraically closed fields. For a particular Q, where Q_{ab} is either finitely generated or a torsion group, it is sufficient for having $\mathrm{Ext}(Q_{ab},A) = 0$ that the elements of A can be divided by all natural numbers which appear as orders of elements of Q_{ab}. (For finite groups Q_{ab}, such groups sometimes are called $|Q_{ab}|$-divisible, cf. YAMAZAKI [1].) Both conditions in 2.12 hold, if A is divisible and contains elements of finite order n, for all n which appear as orders of elements in the quotients of N. This is satisfied if A is divisible and contains elements of arbitrary finite order. In the category of abelian groups, these groups are exactly the injective objects which are cogenerators. Examples are given by $\mathbb{Q}/\mathbb{Z}$, the torus group $\mathbb{R}/\mathbb{Z}$, the multiplicative groups of algebraically closed fields of characteristic 0, i.e. $\mathbb{C}^*$.

Summarizing part of the results above, we obtain:

2.13 PROPOSITION. The following conditions are equivalent:

(i) e is a generalized A-representation group for some divisible abelian group A containing elements of arbitrary finite order.

(ii) e is a generalized A-representation group for all divisible abelian groups A containing elements of arbitrary finite order.

(iii) e is a generalized representation group (case $A = \mathbb{C}^*$).

(iv) $\theta_*(e)$ is a monomorphism.

(v) The inflation $M(\pi): M(G) \to M(Q)$ vanishes.

(vi) All complex projective representations of Q can be lifted in e.

2.14 PROPOSITION. The following conditions are equivalent:

(i) e is an A-representation group for some divisible abelian group A with elements of arbitrary finite order.

(ii) e is an A-representation group for all divisible abelian groups A with elements of arbitrary finite order.

(iii) e is a representation group.

(iv) $\theta_*(e)$ is an isomorphism.

(v) e is a stem extension and $M(\pi) = 0$.

(vi) e is a stem extension and all complex projective representations of Q can be lifted in e.

PROOF of 2.13 and 2.14. The equivalence of (i), (ii), (iii), (iv) is immediate by 2.8, the equivalence of (iv) and (v) follows from I.3.5, and the equivalence of (iii) and (vi) from 2.3. □

2.15 COROLLARY. The following conditions are equivalent:

(i) Each central extension of Q is a generalized representation group.

(ii) The trivial extension $0 \hookrightarrow Q \twoheadrightarrow Q$ is (up to isomorphism) the only representation group of Q.

(iii) $M(Q) = 0$.

(iv) Each complex projective representation of Q is equivalent to a linear representation. □

2.16 COROLLARY. Let e be any group extension (not necessarily central) such that $M(\pi): M(G) \to M(Q)$ vanishes. In particular, this holds whenever $M(G) = 0$. Then the centralization c(e) is a generalized representation group of Q.

PROOF. Let $c(e) = (\cdot, \pi')$. Then $\mathrm{Im}\, M(\pi') = \mathrm{Im}\, M(\pi)$ yields $M(\pi') = 0$. Apply 2.13. □

Now we return to more general abelian groups A.

2.17 DEFINITION. Assume $\mathrm{Ext}(Q_{ab}, A) = 0$. An A-representation group e of Q is called minimal, if it is a stem extension satisfying $\mathrm{Ker}\ \theta_*(e) = L(Q,A)$.

In this terminology, all A-representation groups are minimal, if A is divisible with elements of arbitrary finite order. Now let e be a minimal A-representation group (for an arbitrary A), and $e' \colon N' \rightarrowtail G' \twoheadrightarrow Q$ a generalized A-representation group. Then we have

$N \cong M(Q)/L(Q,A)$ and $\mathrm{Ker}\ \theta_*(e') \subseteq L(Q,A)$.

Hence there exists a monomorphism from N into a quotient of N', which justifies Definition 2,17. If Q is finite, we have by 3.10 that M(Q) is finite, and we obtain:

2.18 PROPOSITION. Let Q be finite and $\mathrm{Ext}(Q_{ab},A) = 0$. Then the minimal A-representation groups of Q are exactly the finite generalized A-representation groups of Q whose middle groups have minimal order. □

Let us summarize some of the results above in the case where A is the multiplicative group of an algebraically closed field K of characteristic $p \geq 0$. Then K* is divisible, and we have $L(Q,K^*) = M(Q)_p$, the group of all p-elements in M(Q), resp. $M(Q)_p = 0$ for p=0. Hence the minimal K*-representation groups of Q are given by the stem extensions e satisfying $\mathrm{Ker}\ \theta_*(e) = M(Q)_p$. If p=0, we obtain the representation groups, and no other K*-representation groups occur, cf. 2.14.

2.19 LEMMA. Let M, N_1, N_2 be abelian groups, A a divisible abelian group, $\theta_i \colon M \to N_i$ homorphisms, and L, resp. L_i the intersection of the kernels of all homomorphisms from M, resp. N_i to A. By $\theta_i{}^*$ we denote the homomorphism from $\mathrm{Hom}(N_i,A)$ to $\mathrm{Hom}(M,A)$ induced by θ_i. Then we have:

(i) $\mathrm{Ker}\ \theta_1 + L = \mathrm{Ker}\ \theta_2 + L$ implies $\mathrm{Im}\ \theta_1{}^* = \mathrm{Im}\ \theta_2{}^*$.

(ii) The converse in (i) holds if one of the following conditions is satisfied:

(a) $L_i \cap \mathrm{Im}\ \theta_i \subseteq \theta_i(L)$, $i=1,2$.

(b) A is divisible with elements of arbitrary finite order.

(c) M is a torsion group.

PROOF. (i) Assume $\mathrm{Ker}\ \theta_1 + L = \mathrm{Ker}\ \theta_2 + L$, and let $\alpha \in \mathrm{Hom}(N_1,A)$. Thus, $\theta_1{}^*(\alpha)=\alpha\theta_1 \in \mathrm{Hom}(M,A)$ and

$\mathrm{Ker}\ \alpha\theta_1 \supseteq \mathrm{Ker}\ \theta_1 + L = \mathrm{Ker}\ \theta_2 + L$.

Then the following homomorphism is well-defined

$\beta' : \mathrm{Im}\ \theta_2 \to A : \theta_2(x) \longmapsto \alpha\theta_1(x)$.

As A is divisible, we have an extension $\beta: N_2 \to A$ of β'. From the definition follows $\theta_2{}^*(\beta) = \beta\theta_2 = \theta_2{}^*(\alpha)$.

(ii) Assume $\mathrm{Im}\ \theta_1{}^* = \mathrm{Im}\ \theta_2{}^*$ and $L_i \cap \mathrm{Im}\ \theta_i \subseteq \theta_i(L)$, $i=1,2$, and let $x \in \mathrm{Ker}\ \theta_1 + L$. Thus we have $\alpha\theta_1(x) = 0$ for all $\alpha \in \mathrm{Hom}(N_1,A)$. From $\mathrm{Im}\ \theta_1{}^* = \mathrm{Im}\ \theta_2{}^*$ we obtain $\beta\theta_2(x) = 0$ for all $\beta \in \mathrm{Hom}(N_2,A)$. This implies $\theta_2(x) \in \mathrm{Im}\ \theta_2 \cap L_2 \subseteq \theta_2(L)$, and we obtain $x \in \mathrm{Ker}\ \theta_2 + L$, finishing case (a).

If A is divisible with elements of arbitrary finite order we have $L = L_i = 0$. In general L, resp. L_i consists of all torsion elements of M, resp. N_i, whose orders are divisible only by those primes p for which no elements of order p exist in A. If M is a torsion group we therefore have $\mathrm{Im}\ \theta_i \cap L_i = \theta_i(L)$. Thus, cases (b) and (c) are contained in (a). □

For later purposes we need:

2.20 PROPOSITION. Let K be an algebraically closed field of characteristic p, $e_i\colon N_i \rightarrowtail G_i \twoheadrightarrow Q$, i=1,2 , central extensions, and assume that either M(Q) is a torsion group or p = 0 . Then the same projective K-representations of Q can be lifted in e_1 and e_2 , if and only if

$$\mathrm{Ker}\ \theta_*(e_1) + M(Q)_p = \mathrm{Ker}\ \theta_*(e_2) + M(Q)_p \ .$$

PROOF. By 2.19 the equation above is equivalent to $\mathrm{Im}\ \theta_*(e_1)^* = \mathrm{Im}\ \theta_*(e_2)^*$, and from 2.5 (ii), 2.6 and 1.6 this holds, if and only if the same projective representations can be lifted. □

The connection between various situations occurring above and some of the results are shown by the following diagrams:

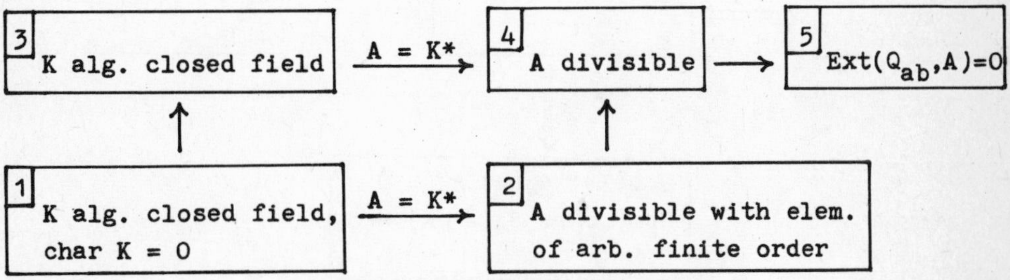

In the two following diagrams, the numbers beside the arrows refer to the properties of A numbered in the first diagram. The boxes contain the numbers of the corresponding definitions and propositions.

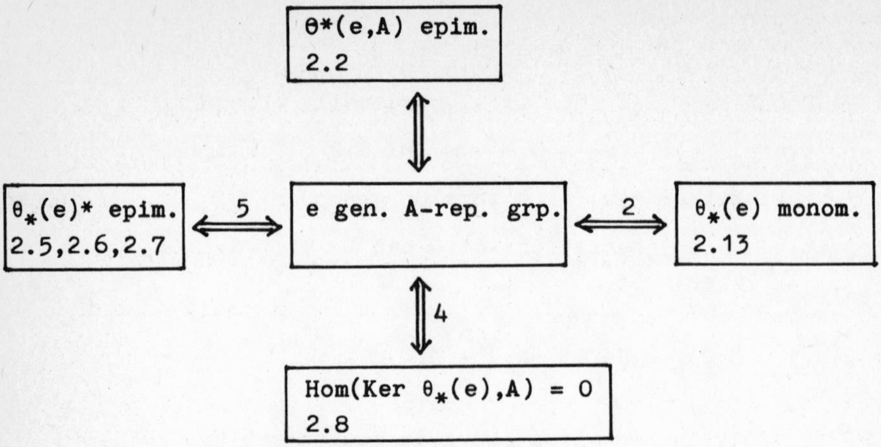
θ*(e,A) epim.
2.2
θ*(e)* epim.
2.5,2.6,2.7
5
e gen. A-rep. grp.
2
θ*(e) monom.
2.13
4
Hom(Ker θ*(e),A) = 0
2.8

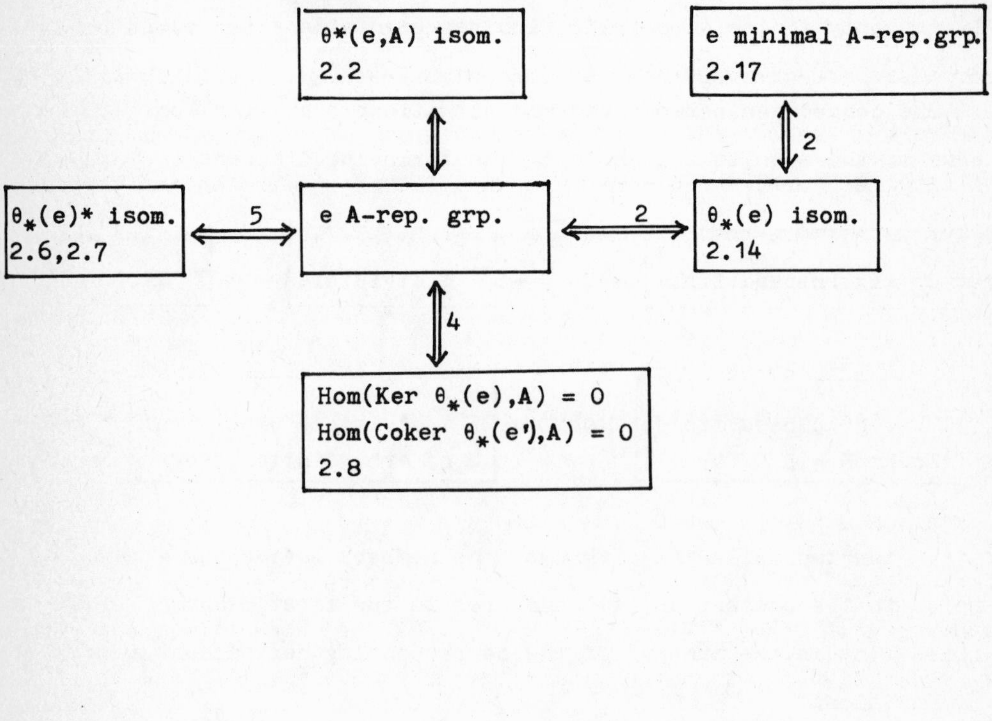
θ*(e,A) isom.
2.2
e minimal A-rep.grp.
2.17
2
θ*(e)* isom.
2.6,2.7
5
e A-rep. grp.
2
θ*(e) isom.
2.14
4
Hom(Ker θ*(e),A) = 0
Hom(Coker θ*(e'),A) = 0
2.8

3. Representation Groups

In this section we restrict ourselves to projective representations over algebraically closed fields K. Certain K*-representation groups for Q are given by the central extensions e by Q with $\theta_*(e)$ isomorphic. These extensions were called the representation groups of Q in Section 2, and they proved to be the only ones if char K = 0 .

3.1 DEFINITION. (i) Let e_1 and e_2 be group extensions by Q. A morphism of the form $(\alpha,\beta,1_Q): e_1 \to e_2$ is called a homomorphism over Q, and if β is isomorphic, an isomorphism over Q. These homomorphisms are the morphisms in the extension category $(\frac{Q}{})$ in GRUENBERG [1; chp. 9].

(ii) Let $e: N \overset{\kappa}{\rightarrowtail} G \overset{\pi}{\twoheadrightarrow} Q$ be an arbitrary extension and U a subgroup of N, such that κU is normal in G. Then e/U denotes the induced extension $N/U \overset{\kappa'}{\rightarrowtail} G/\kappa U \overset{\pi'}{\twoheadrightarrow} Q$.

Now we are going to describe the representation groups of Q in terms of isomorphism over Q. Let $e: N \overset{\kappa}{\rightarrowtail} G \twoheadrightarrow Q$ be a representation group of Q, and

$$e' : M(Q) \overset{\lambda}{\rightarrowtail} G \twoheadrightarrow Q \quad , \lambda = \kappa\theta_*(e)$$

Then $\theta_*(e') = 1_{M(Q)}$, and e and e' are isomorphic over Q. On the other hand, let $e_i: N \rightarrowtail G_i \twoheadrightarrow Q$, i=1,2 be stem extensions with $\theta_*(e_1) = \theta_*(e_2)$, then the naturality of I.(3.4) yields $\alpha = 1_N$, and β is isomorphic, whenever $(\alpha,\beta,1_Q): e_1 \to e_2$ is a homomorphism over Q, i.e. $[e_1] = [e_2]$. If we combine these observations with I.3.8, we obtain:

3.2 PROPOSITION. (i) Each representation group of Q is isomorphic over Q to a central extension $e': M(Q) \rightarrowtail G \twoheadrightarrow Q$ with $\theta_*(e') = 1_{M(Q)}$.

(ii) The classes of representation groups of Q - with respect to isomorphism over Q - are uniquely parametrized by their members in $\theta_*^{-1}(1_{M(Q)})$, which is a coset of $\Psi\,\mathrm{Ext}(Q_{ab},M(Q))$ in $\mathrm{Cext}(Q,M(Q))$. □

3.3 REMARK. The classes of representation groups under arbitrary isomorphism of extensions will be considered in Chapter III. One should keep in mind that isomorphism of extensions is a much weaker condition than isomorphism over Q as in 3.2. In group theory it is often important to distinguish only between non-isomorphic middle groups of representation groups, a condition even weaker than isomorphism of extensions. Hence, the size of $\mathrm{Ext}(Q_{ab},M(Q))$ is an upper bound for the number of these groups. If Q is finite, M(Q) is also finite by 3.10,(i). Let $a_1,a_2,\ldots,a_n$ and $b_1,\ldots,b_m$ be the invariants of Q_{ab} and M(Q) respectively. Then the greatest common divisors (a_i,b_j) , $1 \leq i \leq n$, $1 \leq j \leq m$ are the invariants of $\mathrm{Ext}(Q_{ab},M(Q))$, cf. III.4.5. This bound for the number of representation groups of finite groups is due to SCHUR [2; Satz I, p.95].

As already mentioned in Section 2, the Universal Coefficient Theorem yields the existence of representation groups. The following proposition, essentially due to Schur, gives a more explicit description.

3.4 PROPOSITION, cf. SCHUR [2; §3], GRUENBERG [1, chp. 9.9].
(i) Let $e(F,R): R \hookrightarrow F \twoheadrightarrow Q$ be a free presentation of Q, $S \subseteq R$ such that $S/[R,F]$ is a complement of $(R\cap[F,F])/[R,F]$ in $R/[R,F]$. Then $e(F,R)/S$ is a representation group of Q.

(ii) For each representation group e of Q there exists a normal subgroup S of F, contained in R, such that $e(R,F)/S$ is isomorphic over Q to e; for each such S, the group $S/[R,F]$ has the property of (i).

(iii) Let e' be a stem extension of Q. Then there exists a representation group e of Q and a subgroup U in the kernel of e, such that e/U and e' are isomorphic over Q.

PROOF. Let $e': N \rightarrowtail G \twoheadrightarrow Q$ be a stem extension. The identity of Q gives rise to homomorphisms $\alpha: R \to N$, $\beta: F \to G$, such that diagram I.(3.6) is commutative. For sake of simplicity we evaluate $M(Q)$ at $e(F,R)$, i.e. $M(Q) = (R\cap[F,F])/[R,F]$. Then α induces the epimorphism $\theta_*(e'): M(Q) \twoheadrightarrow N$, which implies that α and β are epimorphisms, and we put $T := \operatorname{Ker}\alpha = \operatorname{Ker}\beta$, which satisfies $[R,F] \subseteq T \subseteq R$ and $T\cdot(R\cap[F,F]) = R$. This yields

$$T/(T\cap[F,F]) \cong R/(R\cap[F,F]) \cong R[F,F]/[F,F] ,$$

and $T/(T\cap[F,F])$ is free abelian. Let S be a subgroup of T such that $S/[R,F]$ is a complement of $(T\cap[F,F]/[R,F])$ in $T/[R,F]$, which implies

$$S \cap [F,F] = S \cap (R\cap[F,F]) = S \cap (T\cap[F,F]) = [R,F] ,$$

i.e. S satisfies the conditions in (i). We also have the following obvious isomorphisms over Q:

$$e' \cong e(F,R)/T \cong (e(F,R)/S)/(T/S) .$$

We have

$$\theta_*(e(F,R)/S) = (x[R,F] \longmapsto xS) , \; x \in R\cap[F,F]$$

$$\theta_*(e(F,R)/T) = (x[R,F] \longmapsto xT) , \; x \in R\cap[F,F] .$$

As $\theta_*(e(F,R)/S)$ is an obvious isomorphism, $e(F,R)/S$ is a representation group, and it follows that e' is a representation group, if and only if $T = S$. These observations easily prove (i), (ii) and (iii). □

3.5 REMARK. Let us consider the extensions $e(F,R)/S$ from 3.4. The isomorphisms $\theta_*(e(F,R)/S)\colon M(Q) \rightarrowtail\!\!\twoheadrightarrow R/S$, cf. proof of 3.4, allow us to identify $M(Q)$ and R/S. Thus we regard the representation groups $e(F,R)/S$ as those central extensions e by Q for which $\theta_*(e)$ is the identity, and from 3.4 (ii),(iii) we obtain that each stem extension of Q is forward induced from such an extension e.

A more detailed description of this situation will follow in Section III.4, where presentations of isoclinic groups are studied.

3.6 PROPOSITION. Assume that Q has a presentation with n generators and r relations, and let k be the rank of Q_{ab}/T, where T is the torsion subgroup of Q_{ab}. Then $M(Q)$ can be generated by k+r-n elements. (This proposition will be discussed in more detail in Section IV.1 as P. Hall's Inequality.)

SKETCH OF PROOF. Let $R \hookrightarrow F \twoheadrightarrow Q$ be a free presentation of Q, where F is free of rank n and R the normal closure of r elements. Then $R/[R,F]$ is an abelian group with of most r generators. On the other hand, $R/(R\cap[F,F]$ is free abelian of rank n-k. □

The following proposition is an easy consequence of 2.16, resp. 2.14 and sometimes quite useful in order to exhibit representation groups.

3.7 PROPOSITION. Let $e\colon N \rightarrowtail G \twoheadrightarrow Q$ be a stem extension, and assume $M(G) = 0$. Then e is a representation group of Q. □

3.8 EXAMPLES OF REPRESENTATION GROUPS. (i) Groups with trivial multiplicator are their own representation groups. In particular, by Proposition 3.6, this covers finite groups which can be presented with the same number of generators and relations. We mention in passing that cyclic and free groups have trivial multiplicator by Lemma I.3.4.

(ii) The generalized quaternion group of order 2^{n+1} has a presentation $R \hookrightarrow F \twoheadrightarrow Q$, where F is free on $\{x,y\}$, and R the normal closure of $\{x^{2^{n-1}}y^2, x^2[y^{-1},x^{-1}]\}$. Hence we can apply 3.6 in order to see that the multiplicator vanishes.

(iii) Let Q be abelian on two generators. In Example I.4.6(i), (ii) an extension e by Q is given with $\theta_*(e)$ isomorphic. Hence it is a representation group. Let us consider the group $Q = \mathbb{Z}/p \times \mathbb{Z}/p$ more closely, where p is a prime. From I.4.6 it follows $M(Q) \cong \mathbb{Z}/p$. Now we have the presentation $R \hookrightarrow F \twoheadrightarrow Q$, where F is free on x,y, and

$$R = \langle x^p, y^p, [x,y]\rangle^F ,$$

the upper index F denotes normal closure. Thus we have

$$R \cap [F,F] = [F,F] \text{ , and } [R,F] = \langle [x,y]^p, [F,F,F]\rangle^F ,$$

and the complements of $[F,F]/[R,F]$ in $R/[R,F]$ are given by $C_{n,m}/[R,F]$, where

$$C_{n,m} = \langle x^p[x,y]^n, y^p[x,y]^m, [R,F]\rangle^F , \quad o \leq n,\ m \leq p-1 .$$

Hence we have p^2 complements. On the other hand $\mathrm{Ext}(Q_{ab}, M(Q)) \cong \mathrm{Ext}(\mathbb{Z}/p \times \mathbb{Z}/p, \mathbb{Z}/p)$ has also the order p^2. By 3.2 and 3.5, the central extensions

$$R/C_{n,m} \hookrightarrow F/C_{n,m} \twoheadrightarrow Q$$

are a system of representatives of representation groups of Q with respect to isomorphism over Q, see also III.4.7.

(iv) The dihedral group D of order 2^{n+1} has the following presentation $R \hookrightarrow F \twoheadrightarrow D$, F free on x,y and

$$R = \langle x^2, [x,y]y^2, [x,y]^{2^{n-1}} \rangle^F ,$$

and let $G = F/\langle x^2[x,y]^{-2}, [x,y]y^2, [x,y]^{2^n} \rangle^F$. Then G is isomorphic to the generalized quaternion group of order 2^{n+2} and gives rise to a stem extension

$$\mathbb{Z}/2 \rightarrowtail G \twoheadrightarrow D .$$

By (ii), G has trivial multiplicator, and by 3.7 the extension above is a representation group of D, in particular we obtain that $[x,y]^{2^{n-1}}[R,F]$ is a generator of $M(D) = (R \cap [F,F])/[R,F] \cong \mathbb{Z}/2$. The elements $x^2[R,F]$, $[x,y]y^2[R,F]$ generate a complement of $M(D)$ in $R/[R,F]$. Hence we have exactly $4 = |\mathrm{Ext}(D_{ab}, M(D))|$ complements $C_{ij}/[R,F]$, $0 \leq i,j \leq 1$ given by

$$C_{ij} = \langle x^2[x,y]^{2^{n+i-1}}, [x,y]^{2^{n+j-1}+1}y^2, [R,F] \rangle^F .$$

The same argument as in example (iii) shows that the extensions

$$R/C_{ij} \hookrightarrow F/C_{ij} \twoheadrightarrow D$$

are representatives of the representation groups under isomorphism over D. The group F/C_{ij} is a dihedral group if $i=j=1$, a generalized quaternion group if $i=0$, $j=1$ and a quasi-dihedral group if $j=0$, $i=0$ or 1 .

(v) The representation groups of the symmetric and alternating groups were studied by de SEGUIER [1] and SCHUR [3], and SCHUR [2] obtained the representation groups of $SL(2,q)$ and $PSL(2,q)$, see also HUPPERT [1; V.25].

(vi) Representation groups of perfect and of metacyclic groups will be discussed in II.5 and IV.2, respectively.

3.9 PROPOSITION (Schur). Let K be an arbitrary field and $\gamma: Q \to PGL(V)$ a projective K-representation of finite dimension n. Then the associated cohomology element $\gamma^*(\sigma_V) \in \mathrm{Cext}(Q,K^*)$ has finite order dividing n.

PROOF. The extension $\gamma^*(\sigma_V)$ is defined by the top row of diagram (1.3). Let $\alpha: K^* \to K^* : x \longmapsto x^n$, and $d=\det\circ\beta: G \to K^*$. Then the definition of the addition in $\mathrm{Cext}(Q,K^*)$, cf. I.2.4, yields $n\cdot\gamma^*(\sigma_V) = \alpha_*\gamma^*(\sigma_V)$. On the other hand we have $\alpha = d\circ\lambda$, and I.1.3 implies $\alpha_*\gamma^*(\sigma_V) = 0$. $\square$

3.10 COROLLARIES (Schur). Let Q be a finite group. Then:

(i) M(Q) is a finite abelian group.

(ii) If K is an algebraically closed field of characteristic 0, then $\mathrm{Cext}(Q,K^*) \cong H^2(Q,K^*) \cong M(Q)$.

(iii) If $R \hookrightarrow F \twoheadrightarrow Q$ is a free presentation of Q, then M(Q) is the torsion subgroup of $R/[R,F]$.

(iv) Let e be the exponent of M(Q). Then e^2 divides the order of Q.

REMARK. 3.10 (ii) usually does not hold if Q is infinite. An example is given by $Q = \mathbb{Z} \times \mathbb{Z}$. Here we have $M(Q) \cong \mathbb{Z}$ by I.4.6 , while I.3.8 yields $\mathrm{Cext}(Q,K^*) \cong H^2(Q,K^*) \cong K^*$.

PROOF of 3.10. (i) By 3.6, M(Q) is finitely generated. Let K be an algebraically closed field of characteristic 0. Then $\mathrm{Cext}(Q,K^*) \cong \mathrm{Hom}(M(Q),K^*)$ by I.3.8, and from 1.6 and 3.9 it follows that $\mathrm{Hom}(M(Q),K^*)$ is a torsion group. Thus M(Q) is a torsion group as well, and M(Q) is finite.

(ii) The assertion follows from the proof of (i).

(iii) As mentioned above, $R/(R\cap[F,F])$ is a free abelian group, and by (i), $M(Q) = (R\cap[F,F])/[R,F]$ is finite. This proves (iii).

(iv) Let K be as above, and $x \in \mathrm{Cext}(Q,K^*)$. As Q is finite, we have by (i) a finite representation group e of Q:

$$e : M(Q) \overset{\varkappa}{\rightarrowtail} G \twoheadrightarrow Q ,$$

and there is a (unique) $\alpha \in \mathrm{Hom}(M(Q),K^*)$ satisfying $\theta^*(e,K^*)\alpha = x$. Let $\beta_1,\beta_2,\ldots,\beta_n$ be the irreducible constituents of the induced representation $(\alpha\varkappa^{-1})^G$. Then each β_i gives rise to a projective representation of Q having cohomology element x. The Frobenius Reciprocity Theorem yields that β_i is contained in $(\alpha\varkappa^{-1})^G$ exactly $(\dim\beta_i)$-times and we obtain

$$|Q| = \dim(\alpha\varkappa^{-1})^G = \sum(\dim\beta_i)^2 .$$

This equation and 3.9 show that the square of the order of x divides $|Q|$, and (iv) follows from (ii). □

The theory of representation groups can be used to study projective representations of finite groups by applying the theory of linear representations. A nice example is the following theorem of FRUCHT [1]:

3.11 THEOREM. Let Q be a finite abelian group, K an algebraically closed field and $x \in \mathrm{Cext}(Q,K^*)$. Then any two irreducible projective representations of Q with cohomology element x are projectively equivalent.

PROOF. Let $N \rightarrowtail G \twoheadrightarrow Q$ be a minimal K^*-representation group for Q and $p = \mathrm{char}(K)$. Then N is isomorphic to the p'-part of M(Q). Without loss of generality we can assume $N \subseteq G$, and we have $G = G' \times Q_p$, where Q_p is isomorphic to the p-part of Q, and G' is a nilpotent p'-group of class at most 2, containing N. By 1.10 we have to show that any two irreducible linear representations β_1, β_2 of G', whose restrictions to N contain the same (one-dimensional) constituent α are equivalent up to a one-dimensional factor. As G' is an M-group, we have subgroups A_1, A_2 satisfying $A_1 \cap A_2 \supseteq Z(G') \supseteq N$, and one-dimensional representations α_i of A_i with $\alpha_i^{G'} = \beta_i$, which thus satisfy $\alpha_i|_N = \alpha$. Hence,

$\sigma_1 := \alpha_1\alpha_2^{-1}$ can be regarded as a character of $(A_1 \cap A_2)/N$, and let σ_2 resp. σ_3 be extensions of σ_1 to A_2/N resp. G'/N. Now we have $\sigma_3\beta_2 = (\sigma_2\alpha_2)^{G'}$. As $\sigma_2\alpha_2$ and α_1 coincide on $A_1 \cap A_2$, we obtain by Mackey's induction theorem that β_1 and $\sigma_3\beta_2$ have a non-trivial common constituent. As they are irreducible, they are equivalent. □

Frucht's theorem can be generalized to arbitrary finite groups, see MANGOLD [1], TAPPE [2].

3.12 EXAMPLE. Let Q be an elementary abelian group of order p^{2n}, p a prime, and consider a stem extension $\mathbb{Z}/p \rightarrowtail E \twoheadrightarrow Q$, where E is extra-special. In HUPPERT [1;V,16.14] it is shown that the ordinary irreducible representations of E of dimension greater 1 all have dimension p^n. The corresponding irreducible projective representations of Q have cohomology elements of order p, because they are derived from irreducible characters of $\mathbb{Z}/p$. Using Frucht's theorem or just the fact that the irreducible representations (beside the one-dimensional ones) have dimension p^n, we obtain that there cannot exist projective representations of lower dimension than p^n, having the cohomology elements from above. Hence the bound for the order of the cohomology elements given by Proposition 3.9, is sometimes very bad.

3.13 REMARK. Another application of lifting projective representations has been given by PAHLINGS [1;4.1]. Here the finite groups with faithful irreducible projective representations are characterized. The result reads as follows: Let K be an algebraically closed field of characteristic p. Then the finite group G has a faithful irreducible projective K-representation, if and only if G has no normal p-subgroup and is the central factor group of a finite group H, whose socle is the normal closure of a single element.

We close this section with a few remarks on unitary projective representations. Let H be a Hilbert space with inner product $\langle\ ,\ \rangle$. By U(H) we denote the subgroup of GL(H) that consists of all u satisfying $\langle ux,uy\rangle = \langle x,y\rangle$ for all $x,y \in H$, and let T denote the multiplicative group of all complex numbers of absolute value 1. Then we have the central extension

$$\tau_H :\ T \overset{\delta_H}{\rightarrowtail} U(H) \twoheadrightarrow PU(H)\ ,$$

where $\delta_H(t) = (x \longmapsto tx)$ and $PU(H) = \mathrm{Coker}\ \delta_H$. Any homomorphism $\gamma\colon Q \to PU(H)$ is called a unitary projective representation of Q, and $\gamma^*(\tau_H) \in \mathrm{Cext}(Q,T)$ the associated cohomology element. A similar argument as in 1.6 shows that each element of Cext(Q,T) is induced by a unitary projective representation over the Hilbert space $l^2_{|Q|}$. As $T \cong \mathbb{R}/\mathbb{Z}$ is divisible with elements of arbitrary finite order, the question of lifting unitary projective representations also leads to the representation groups of Q by the results of Section 2.

4. Representation Groups of Free and Direct Products

In this section we construct (generalized) representation groups for the free product $Q_1 * Q_2$ and the direct product $Q_1 \times Q_2$ of two groups from central extensions

(4.1) $\quad e_k = (\varkappa_k, \pi_k) : N_k \rightarrowtail G_k \twoheadrightarrow Q_k$, $k=1,2$,

which are (generalized) representation groups of Q_k according to the context. In homological terms, the resulting formulas for $M(Q_1 * Q_2)$ and $M(Q_1 \times Q_2)$ are well-known; a different proof of these formulas, also non-homological, was given by C.MILLER [1]. Concerning the direct product, the basic idea is taken from SCHUR [2] and WIEGOLD [2]. Since we here allow infinite groups, the proofs need to look different. Actually, SCHUR [2;Satz VI,p.109] described $M(Q_1 \times Q_2)$ for finite groups Q_1 and Q_2.

4.1. Recall that the category of groups has sums, these are traditionally called <u>free products</u>. In other words, given any groups G_1 and G_2, one has a group $G_1 * G_2$ and homomorphisms $j_k: G_k \to G_1 * G_2$ for $k=1,2$ with the following universal property: For every pair of homomorphisms $f_K: G_k \to H$

(4.2)

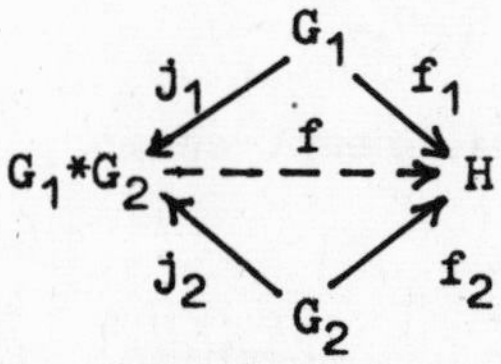

with common range, there is a unique homomorphism $f = \{f_1, f_2\}: G_1 * G_2 \to H$ with $f \circ j_k = f_k$ for $k=1,2$. We also need the projections $q_1 = \{1,0\}: G_1 * G_2 \to G_1$ and $q_2 = \{0,1\}: G_1 * G_2 \to G_2$ which by the very definition satisfy

$$(4.3)\quad \begin{aligned} &q_k \circ j_k = 1 = \text{Identity von } G_k \quad \text{for } k=1,2 \; ; \\ &q_2 \circ j_1 = 0 : \; G_1 \to G_2 \; , \; q_1 \circ j_2 = 0 : \; G_2 \to G_1 \; . \end{aligned}$$

Free products exist for arbitrary (possibly infinite) families of groups, but our main concern is with two free factors.

4.2 We infer the following explicit construction of the free product. The group $G_1 * G_2$ is defined as F/R where F is the free group on the generators $\bar{a}$ and $\bar{b}$, one $\bar{a}$ for each $a \in G_1$ and one $\bar{b}$ for each $b \in G_2$, and R is the normal subgroup of F generated by $\bar{a}_1 \cdot \bar{a}_2 \cdot (\overline{a_1 \cdot a_2})^{-1}$ and $\bar{b}_1 \cdot \bar{b}_2 \cdot (\overline{b_1 \cdot b_2})^{-1}$. The homomorphisms j_1 and j_2 are defined by $j_1(a) = \bar{a}$ and $j_2(b) = \bar{b}$ for $a \in G_1$ and $b \in G_2$. It turns out that every element in $G_1 * G_2$ has a unique <u>reduced</u> or <u>normal form</u> as 1 or an "alternating product", for example $\bar{b}_1 \bar{a}_2 \bar{b}_2 \ldots \bar{a}_{2n-1} b_{2n-1} \bar{a}_{2n}$ with $a_i \in G_1 \backslash 0$, $b_i \in G_2 \backslash 0$. The homomorphisms j_1 and j_2 are monomorphic. If the context permits, one often regards G_1 and G_2 as subgroups of $G_1 * G_2$, suppressing j_1 and j_2 and the bars from the notation.

4.3 LEMMA. (a) The free product is a functor from pairs of groups to groups; given $f_k \colon G_k \to H_k$ for k=1,2, then

$$f_1 * f_2 = | j_1 \circ f_1, j_2 \circ f_2 | \colon G_1 * G_2 \to H_1 * H_2 \; .$$

(b) If f_1 and f_2 are monomorphic resp. epimorphic, then so is $f_1 * f_2$.

(c) If f_1 and f_2 are arbitrary homomorphisms, then $\operatorname{Ker}(f_1 * f_2)$ is generated by $j_1(\operatorname{Ker} f_1)$ and $j_2(\operatorname{Ker} f_2)$ as a normal subgroup of $G_1 * G_2$.

(d) If $G_k = F_k / R_k$ are free presentations of G_1 and G_2, where F_k is the free group on the set X_k, then $G_1 * G_2 \cong F/R$ with $F \cong F_1 * F_2$ the free group on the disjoint $X_1 \dot{\cup} X_2$ and R the normal subgroup generated by $j_1(R_1)$ and $j_2(R_2)$.

PROOF. Part (b) is immediate from 4.2 and (d) from (c). Concerning (c), this is the salient point: if an alternating product z lies in $\mathrm{Ker}(f_1 * f_2)$, one must have $f_1(a_k) = 1$ or $f_2(b_k) = 1$ for at least one k, by appeal to the reduced form in $H_1 * H_2$. One proceeds by induction on the "length" of z. □

4.4 THEOREM. Given generalized representation groups e_1 of Q_1 and e_2 of Q_2 as in (4.1). Let $K = \mathrm{Ker}(\pi_1 * \pi_2)$ and consider

$$e = e_1 * e_2 \ : \ K \hookrightarrow G_1 * G_2 \xrightarrow{\pi_1 * \pi_2} Q_1 * Q_2 \ .$$

(a) Then c(e) is a generalized representation group of $Q_1 * Q_2$.

(b) If moreover e_1 and e_2 are representation groups of Q_1 and Q_2, respectively, then c(e) is a representation group of $Q_1 * Q_2$.

(c) $M(Q_1 * Q_2)$ is the internal direct sum of $\mathrm{Im}\, M(j_1)$ and $\mathrm{Im}\, M(j_2)$, where $M(j_k): M(Q_k) \to M(Q_1 * Q_2)$ are monomorphic; thus $M(Q_1 * Q_2) \cong M(Q_1) \times M(Q_2)$. This direct sum decomposition is natural with respect to pairs of homomorphisms $Q_1 \to \bar{Q}_1$, $Q_2 \to \bar{Q}_2$.

PROOF. By Lemma 4.3, $\pi_1 * \pi_2$ is epimorphic and K is generated by $j_1 \varkappa_1 N_1$ and $j_2 \varkappa_2 N_2$ as a normal subgroup of $G_1 * G_2$. In general, K is not central.

(a) Given any complex vector space V and a homomorphism $\gamma: Q_1 * Q_2 \to PGL(V)$. Since the e_k are generalized representation groups, there are homomorphisms $\beta_k: G_k \to GL(V)$ with $\tau\beta_k = (\gamma j_k)\pi_k$,

$$\begin{array}{ccccc} & & G_k & \xrightarrow{\pi_k} & Q_k \\ & & \downarrow{\scriptstyle \beta_k} & & \downarrow{\scriptstyle \gamma \circ j_k} \\ \sigma_V \ : & \mathbb{C}^* \xrightarrow{\ \delta\ } & GL(V) & \xrightarrow{\ \tau\ } & PGL(V) \ . \end{array}$$

Let $\beta = \{\beta_1, \beta_2\}: G_1 * G_2 \to GL(V)$, then clearly $\tau\beta = \gamma \circ (\pi_1 * \pi_2)$.

Thus we obtain a morphism $(\cdot,\beta,\gamma)\colon e \to \sigma_V$ of extensions and, by centralizing, $(\cdot,\beta',\gamma)\colon c(e) \to c(\sigma_V)=\sigma_V$. Since all γ can be lifted in this fashion, $c(e)$ is a generalized representation group of Q_1*Q_2 by Remark 2.3.

(b) We show that $c(e)$ is stem provided e_1 and e_2 are stem. Since the kernel group of $c(e)$ is $K/[K,G]$ with $G = G_1*G_2$, it suffices to prove $K \subseteq [G,G]$. Now $j_k\varkappa_k N_k \subseteq j_k[G_k,G_k] \subseteq [G,G]$ for $k=1,2$ by assumption; use the initial remark.

(c) We analyze the situation of (b) in greater detail. Again, as K is generated by $j_1\varkappa_1N_1$ and $j_2\varkappa_2N_2$ as a normal subgroup, $K/[K,G]$ is generated by α_1N_1 and α_2N_2 as an abelian group, where $\alpha_k\colon N_k \to K/[K,G)$ are the obvious maps. By the definition of $\pi_1*\pi_2$, we have morphisms $(\cdot,j_k,j_k)\colon e_k \to e$ of extensions. The naturality of θ_* yields commutative diagrams

$$\begin{array}{ccc} M(Q_k) & \xrightarrow{\ \theta_*(e_k)\ } & N_k \\ \big\downarrow{\scriptstyle M(j_k)} & & \big\downarrow{\scriptstyle \alpha_k} \\ M(Q_1*Q_2) & \xrightarrow{\ \theta_*(e)\ } & \dfrac{K}{[K,G]} \end{array}$$

for $k=1,2$. The horizontal maps are isomorphisms by assertion (b) combined with Proposition 2.14. Consequently $M(Q_1*Q_2) = M_1\cdot M_2$ with $M_k := \operatorname{Im} M(j_k)$. Now $p_2\circ j_2 = 1$ implies $M(p_2)\circ M(j_2) = 1$, thus $M(j_2)$ is monomorphic. Next, $p_2\circ j_1 = 0$ and $M(0) = 0$ imply $M_1 \subseteq \operatorname{Ker} M(p_2)$. Therefore $M_1 \cap M_2 \subseteq (\operatorname{Ker} M(p_2)) \cap M_2 = 0$. The naturality of the direct sum decomposition is immediate from the functoriality of the free product. $\square$

4.5 PROPOSITION. Let G be the free product of groups G_i, $i \in I$, with $M(G_i) = 0$. Then $M(G) = 0$.

PROOF. By 3.8(i), every complex projective representation of G_i

can be lifted to a linear representation. This property carries over to G by the universal property of the free product. Thus $M(G) = 0$ by Remark 2.3. Alternatively, this is immediate from Theorem 4.4 (c) for finite I and in the general case follows by a direct limit argument; consider the directed system of the finite subsets of I and apply Proposition I.5.10. □

4.6 EXAMPLES. (a) The modular group $PSL(2,\mathbb{Z}) = SL(2,\mathbb{Z})/\text{Center}$ is known to be isomorphic to $\mathbb{Z}/2*\mathbb{Z}/3$ with generators $a = \pm\begin{pmatrix}0 & -1\\ 1 & 0\end{pmatrix}$ of order 2 and $b = \pm\begin{pmatrix}0 & -1\\ 1 & 1\end{pmatrix}$ of order 3. As $M(\mathbb{Z}/2) = M(\mathbb{Z}/3) = 0$, the multiplicator of $PSL(2,\mathbb{Z})$ is trivial. (See Example IV.1.7 (d) for a different approach to this result.) Consequently, every complex projective representation of $PSL(2,\mathbb{Z})$ can be lifted to a linear representation. In particular, this applies to the defining projective representation, treated as a complex one. Indeed, the assignment

$$a \longmapsto \begin{pmatrix}0 & -i\\ i & 0\end{pmatrix}, \quad b \longmapsto \begin{pmatrix}0 & -\omega\\ \omega & \omega\end{pmatrix} \quad \text{with} \quad \omega = \exp(\frac{\pi i}{3})$$

determines a faithful complex representation of degree 2, i.e. $PSL(2,\mathbb{Z})$ appears as a subgroup of $GL(2,\mathbb{C})$.

(b) The infinite dihedral group is defined as

$$D_\infty = \mathbb{Z} \rtimes (\mathbb{Z}/2) = \text{group}\langle a,y : y^2 = 1 , yay^{-1} = a^{-1}\rangle .$$

It is known that D_∞ is isomorphic to the free product of the subgroups $\langle ay\rangle$ and $\langle y\rangle$ of order 2, therefore $M(D_\infty) = 0$. (Continued in Example IV.1.7 (e).)

We now treat the direct product of two groups. The canonical injection and projection maps, as depicted in

$$(4.4) \qquad G_1 \underset{p_1}{\overset{i_1}{\rightleftarrows}} G_1 \times G_2 \underset{p_2}{\overset{i_2}{\leftrightarrows}} G_2 ,$$

satisfy identities analogous to (4.3). The canonical homomorphism can: $G_1 * G_2 \to G_1 \times G_2$ is given by $p_k \cdot can \cdot j_k = 1$ and $p_k \cdot can \cdot j_l = 0$ for $k \neq l$, $1 \leq k, l \leq 2$. It is epimorphic.

4.7 LEMMA. The kernel of can: $G_1 * G_2 \to G_1 \times G_2$ is $[j_1G_1, j_2G_2]$, usually abbreviated to $[G_1, G_2]$.

Recall that, for subgroups A and B of some group, [A,B] denotes the subgroup generated by all $[a,b] = aba^{-1}b^{-1}$ with $a \in A$ and $b \in B$. The conventions of 4.2 explain the notation $[G_1, G_2]$.

PROOF. Clearly $[G_1, G_2] \subseteq Ker(can)$. By I.4.2 (a) $[G_1, G_2]$ is normal. An appeal to the factor group $G_1 * G_2/[G_1, G_2]$ shows that every element of $G_1 * G_2$ has the form $z = a \cdot b \cdot c$ with $a \in j_1G_1$, $b \in j_2G_2$, $c \in [G_1, G_2]$. Then $can(z) = (a,b)$ and $Ker(can) \subseteq [G_1, G_2]$. □

4.8 DEFINITION. The <u>metabelian product</u> A o B of the groups A and B is $A*B/D$ where $D = [[A,B], A*B]$ is the smallest normal subgroup of A*B such that $[A,B]/D$ is central in $A*B/D$.

The name "metabelian product" was coined by GOLOVIN [2]. This construction (in the case of two factors as above) agrees with the earlier "S-product" of LEVI [1]. Another common name is "second nilpotent product"; this is also due to GOLOVIN [1], exept that he used a different convention for counting the nilpotency class ("first" instead of "second"). Following WIEGOLD [2], we are going to construct a representation group of a direct product as a metabelian product of representation groups.

4.9 PROPOSITION. Let A and B be arbitrary groups. The subgroup [A,B] of A*B is a free group with basis $\{ [a,b] \mid a \in A, b \in B, a \neq 1 \neq b \}$.

This result and the following proof are due to GOLOVIN [1;chp.I].

PROOF. Every element x of $[A,B]$ is a product of some $n \geq 0$ "elementary" commutators, the latter being of the form $[a,b]$ or $[b,a] = [a,b]^{-1}$ with $a \in A$, $b \in B$, $a \neq 1 \neq b$. Without loss of generality, an elementary commutator in x is not followed by its inverse. We prove by induction that, for $n \geq 1$, the last two letters survive unchanged in the unique reduced form of x, as explained in 4.2. The case $n = 1$ is clear. Consider $w_n = \ldots[a_n,b_n]$ which by induction reduces to $w_n = \ldots b^* a_n^{-1} b_n^{-1}$. If $w_{n+1} = w_n[a_{n+1},b_{n+1}]$, no cancellation is possible. If $w_{n+1} = w_n[b_{n+1},a_{n+1}]$ with $b_{n+1} \neq b_n$, then w_{n+1} reduces to $\ldots b^* a_n^{-1}(b_n^{-1}b_{n+1})a_{n+1}b_{n+1}^{-1}a_{n+1}^{-1}$; the last two letters remain untouched. If $w_{n+1} = w_n \cdot [b_n,a_{n+1}]$, then $a_{n+1} \neq a_n$ by our specification. Thus w_{n+1} reduces to $\ldots b^*(a_n^{-1}a_{n+1})b_n^{-1}a_{n+1}$ and the last two letters survive unchanged. The case $w_n = \ldots[b_n,a_n]$ is handled by symmetry. Consequently, $w_n \neq 1$ for all $n \geq 1$ and the set of elementary commutators is a basis. □

4.10 THEOREM (MacHENRY [1], cf. WIEGOLD [1]). For arbitrary groups A and B, there exists a central extension of groups,

$$(4.5) \qquad e : A \otimes B \overset{\varkappa}{\rightarrowtail} A \circ B \overset{\pi}{\twoheadrightarrow} A \times B ,$$

with $\varkappa(\bar{a}\otimes\bar{b}) = [a,b]\cdot D$ and $\pi(aD) = (a,1)$, $\pi(bD) = (1,b)$. Here $A \otimes B$ is defined as $A_{ab} \otimes B_{ab}$ and $\bar{a} = a[A,A]$ etc..

PROOF. It is clear that $\mathrm{can}: A*B \to A \times B$ induces π. Since $[A,B]/D$ is central in $A \circ B$, the set map $(a,b) \longmapsto [a,b]D : A \times B \to A \circ B$ is a bihomomorphism by Lemma I.4.1. This map factors over $A \times B \to A_{ab} \times B_{ab}$, and the universal property of the tensor product yields $\varkappa$ as specified. Now $\mathrm{Ker}\, \pi = [A,B]/D = \mathrm{Im}\, \varkappa$ by Lemma 4.7. We are going to construct $\psi: \mathrm{Im}\, \varkappa \to A\otimes B$

with $\psi\circ\varkappa' = 1$ where $\varkappa' = \varkappa|_{A\otimes B,\mathrm{Im}\,\varkappa}$; consequently $\mathrm{Ker}\,\varkappa = 0$. To this end, we invoke Proposition 4.9 and define a homomorphism $\varphi: [A,B] \to A\otimes B$ by $\varphi[a,b] = \bar{a}\otimes\bar{b}$ for $a \in A$ and $b \in B$. (The formula extends to the cases $a = 1$ and $b = 1$.) Recall the identities I.4.1 (a),(b) in the form

$$^{x}[a,b] = [xa,b]\cdot[x,b]^{-1} , \quad ^{y}[a,b] = [a,y]^{-1}\cdot[a,yb]$$

where now $a,x \in A$ and $b,y \in B$. Thus

$$\varphi(^{x}[a,b]) = (\bar{x}\cdot\bar{a})\otimes\bar{b} - \bar{x}\otimes\bar{b} = \bar{a}\otimes\bar{b} = \varphi[a,b] ,$$

$$\varphi(^{y}[a,b]) = - \bar{a}\otimes\bar{y} + \bar{a}\otimes(\bar{y}\cdot\bar{b}) = \bar{a}\otimes\bar{b} = \varphi[a,b] .$$

Consequently $\varphi(^{z}u) = \varphi(u)$ resp. $\varphi[u,z] = 0$ for all $u \in [A,B]$ and all $z \in A*B$. Thus φ annihilates D and induces $\psi: [A,B]/D \to A\otimes B$ with $\psi([a,b]\cdot D) = \bar{a}\otimes\bar{b}$. Clearly $\psi\varkappa(\bar{a}\otimes\bar{b}) = \bar{a}\otimes\bar{b}$ and thus $\psi\circ\varkappa' = 1$. □

4.11 THEOREM. Given central extensions e_1 by Q_1 and e_2 by Q_2 as in (4.1). In the notation of 4.10 consider the extension

$$(4.6) \qquad \bar{e} : N \hookrightarrow G_1 \circ G_2 \overset{\rho}{\twoheadrightarrow} Q_1 \times Q_2 ,$$

where $\rho=(\pi_1\times\pi_2)\circ\pi : G_1 \circ G_2 \to G_1 \times G_2 \to Q_1 \times Q_2$ and $N = \mathrm{Ker}\,\rho$.

(a) If e_1 and e_2 are generalized representation groups, then $c(\bar{e})$ is a generalized representation group of $Q_1 \times Q_2$.

(b) If both e_1 and e_2 are stem, then so is $\bar{e}$.

(c) If e_1 and e_2 are representation groups, then $\bar{e}$ is a representation group of $Q_1 \times Q_2$.

(d) The Schur multiplicator $M(Q_1\times Q_2)$ is the internal direct sum of $\mathrm{Im}\,M(i_1)$ and $\mathrm{Im}\,M(i_2)$ and $\mathrm{Ker}\,M(p_1) \cap \mathrm{Ker}\,M(p_2) = \mathrm{Im}\,\gamma$, where $\gamma: Q_1\otimes Q_2 \to M(Q_1\times Q_2)$ in terms of (c) is described as the composite map

$$Q_1 \otimes Q_2 \xleftarrow{\pi_1 \otimes \pi_2} G_1 \otimes G_2 \xrightarrow{\varkappa'} \operatorname{Ker} \rho \xleftarrow{\theta_*(\bar{e})} M(Q_1 \times Q_2)$$

with $\varkappa'$ restricting $\varkappa\colon G_1 \otimes G_2 \to G_1 \circ G_2$. There results the Schur-Künneth formula

(4.7) $\quad M(Q_1 \times Q_2) \cong M(Q_1) \times M(Q_2) \times (Q_1 \otimes Q_2)$.

PROOF. (a) We reduce the problem to Theorem 4.4 (a). According to Proposition 2.4, we obtain a generalized representation group of the form

$$\frac{G}{[L,G]} \cong \frac{G/[K,G]}{[L,G]/[K,G]} \xrightarrow{\rho'} Q_1 \times Q_2 ,$$

where $G = G_1 * G_2$, $K = \operatorname{Ker}(\pi_1 * \pi_2)$, $L = (\pi_1 * \pi_2)^{-1}[Q_1,Q_2] = K\cdot[G_1,G_2]$ and ρ' is induced by

$$\mathrm{can}_Q \circ (\pi_1 * \pi_2) = (\pi_1 \times \pi_2) \circ \mathrm{can}_G\colon \ G_1 * G_2 \to Q_1 \times Q_2 .$$

Since $[G_1,G_2]$ is a normal subgroup of G, we have $D = [G,[G_1,G_2]] \subseteq [L,G]$ and $N = L/D$. Consequently, the middle group of $c(\bar{e})$ is

$$\frac{G/D}{[L,G]/D} \cong \frac{G}{[L,G]} .$$

(b) We recall from above that $N = K\cdot[G_1,G_2]/D$. Now $\varkappa_1 N_1 \subseteq [G_1,G_1]$ and $\varkappa_2 N_2 \subseteq [G_2,G_2]$ by assumption. As in the proof of 4.4 (b), we conclude $K \subseteq [G,G]$ and finally $N \subseteq [G,G]/D$. We are left to show that $\bar{e}$ is a central extension. Since D is normal in G, the Three-Subgroups Lemma I.4.3 gives $[\varkappa_1 N_1,G_2] \subseteq [[G_1,G_1],G_2] \subseteq D$. Since $\varkappa_1 N_1$ is central in G_1, we have $[\varkappa_1 N_1,G_1] = 0$ and consequently $[\varkappa_1 N_1,G] \subseteq D$. Similarly $[\varkappa_2 N_2,G] \subseteq D$. By Lemma 4.3 (c), K is generated by $\varkappa_1 N_1$ and $\varkappa_2 N_2$ as a normal subgroup of G. Together, $K\cdot[G_1,G_2]/D$ is central in G/D .

(c) This assertion is the combination of the previous steps;

note that $\bar{e}$ itself is central by (b).

(d) We keep the assumptions of (c). The identities $M(p_1)\cdot M(i_1) = 1$, $M(p_1)\cdot M(i_2) = 0$, etc. exhibit $M(i_k)\colon M(Q_k) \to M(Q_1\times Q_2)$ as monomorphisms and imply

$$M(Q_1\times Q_2) = \operatorname{Im} M(i_1) \times \operatorname{Im} M(i_2) \times \{\operatorname{Ker} M(p_1) \cap \operatorname{Ker} M(p_2)\} ,$$

an internal direct sum. Note that $\pi_1\otimes\pi_2 = (\pi_1)_{ab}\otimes(\pi_2)_{ab}$: $(G_1)_{ab}\otimes(G_2)_{ab} \to (Q_1)_{ab}\otimes(Q_2)_{ab}$ by definition. Under the present assumptions, $(\pi_1)_{ab}$ and $(\pi_2)_{ab}$ and thus $\pi_1\otimes\pi_2$ are isomorphisms. In view of Thm. 4.10, $\varkappa'$ is a well-defined monomorphism. By (c) combined with Proposition 2.14, all of $\theta_*(e_k)$ and $\theta_*(\bar{e})$ are isomorphisms.

We finally derive the formula for $\operatorname{Im} \gamma$. Consider the morphisms $(\beta_k, p_k\cdot\pi, p_k)\colon \bar{e} \to e_k$ of central extensions, where $\beta_k\colon N \to N_k$ denote the restrictions of $p_k\cdot\pi\colon G_1\circ G_2 \to G_1\times G_2 \to G_k$. The naturality of θ_* yields the formulas

$$(4.8) \qquad \beta_k\cdot\theta_*(\bar{e}) = \theta_*(e_k)\cdot M(p_k) \quad \text{for} \quad k=1,2 .$$

Hence $\theta_*(\bar{e}) \operatorname{Im} \gamma \subseteq N \cap \operatorname{Im} \varkappa \subseteq \operatorname{Ker} \beta_1 \cap \operatorname{Ker} \beta_2$ implies $\operatorname{Im} \gamma \subseteq$ $\subseteq \operatorname{Ker} M(p_1) \cap \operatorname{Ker} M(p_2)$. The reverse inclusion follows from (4.8) together with $\operatorname{Ker} \beta_1 \cap \operatorname{Ker} \beta_2 \subseteq \operatorname{Ker}(p_1\pi) \cap \operatorname{Ker}(p_2\pi) = \operatorname{Ker} \pi = \operatorname{Im} \varkappa$.

□

4.12 EXAMPLES. Assume that A and B have trivial multiplicator, thus are their own representation groups. Then Theorem 4.11 exhibits the metabelian product extension (4.5) as a representation group of $A \times B$. Specifically, consider $A = \mathbb{Z}/m$ and $B = \mathbb{Z}/n$ with $n|m$; allowed: $m=0$ or $m=n=0$. Then we obtain

$$(4.9) \qquad G = \text{group} \langle x,y,z : z=[x,y],\ x^m=y^n=1,\ [x,z]=[y,z]=1 \rangle$$

together with the obvious epimorphism $G \to \mathbb{Z}/m \times \mathbb{Z}/n$ as a representation group. In the finite case, these groups were already found by FRUCHT [1; p.19]. We assert that the groups (4.9) are isomorphic to the representation groups described in I.4.6. Indeed, the assignment $x \longmapsto T(1,0,0)\cdot K$, $y \longmapsto T(0,1,0)\cdot K$ is a homomorphism over $\mathbb{Z}/m \times \mathbb{Z}/n$, hence an isomorphism.

4.13 PROPOSITION. Given groups Q_1 and Q_2 with $Q_1 \otimes Q_2 = 0$. If e_1 and e_2 as in (4.1) are representation resp. generalized representation groups, then

$$e_1 \times e_2 : N_1 \times N_2 \rightarrowtail G_1 \times G_2 \twoheadrightarrow Q_1 \times Q_2$$

is a representation resp. generalized representation group of $Q_1 \times Q_2$.

The assumption $Q_1 \otimes Q_2 = 0$ is satisfied whenever $(Q_1)_{ab}$ and $(Q_2)_{ab}$ are torsion groups on disjoint sets of primes, or when Q_1 is perfect. The proposition has an immediate extension to finite direct products and in this form handles finite nilpotent groups; these are the direct product of their Sylow subgroups.

PROOF. Clearly $e_1 \times e_2$ is a central extension, it is stem whenever both e_1 and e_2 are stem. It suffices to prove that $e_1 \times e_2$ is a generalized representation group. The obvious morphisms $(p_k,p_k,p_k)\colon e_1\times e_2 \to e_k$ of extensions yield $p_k \circ \theta_*(e_1\times e_2) = \theta_*(e_k)\circ M(p_k)$ for $k=1,2$. Hence we obtain a commutative diagram

$$\begin{array}{ccc} M(Q_1\times Q_2) & \xrightarrow{\ \theta_*(e_1\times e_2)\ } & N_1 \times N_2 \\ {\scriptstyle \{M(p_1),M(p_2)\}}\downarrow & & \Vert\ {\scriptstyle \{p_1,p_2\}} \\ M(Q_1) \times M(Q_2) & \xrightarrow{\ \theta_*(e_1)\times\theta_*(e_2)\ } & N_1 \times N_2 \end{array}$$

where $\theta_*(e_1) \times \theta_*(e_2)$ is monomorphic by assumption. Moreover, by 4.11 (d) and the present assumption,

$$\mathrm{Ker}\{M(p_1),M(p_2)\} = \mathrm{Ker}\, M(p_1) \cap \mathrm{Ker}\, M(p_2) \cong Q_1 \otimes Q_2 = 0 .$$

Consequently the composite map and finally $\theta_*(e_1\times e_2)$ are monomorphisms. □

The free and direct products are the extreme cases of regular products in the sense of GOLOVIN [1]. We remark that HAEBICH [1] obtained the Schur multiplicators of arbitrary regular products G and, when G is finite, representation groups of G. (In the case of two factors A and B, such a regular product is of the form $A*B/T$ with $T \subseteq [A,B]$. Representation groups of $A*B/T$ can always be obtained in the spirit of our proof of Theorem 4.11.) We finally mention that WIEGOLD [3] and ECKMANN/HILTON/STAMMBACH [2] have treated the multiplicator of central products by arguments related to those given above.

5. The Covering Theory of Perfect Groups

STEINBERG [1] presented a new approach to the problem of how to find free presentations of certain matrix groups. The matrix groups in question were perfect, and KERVAIRE [1] found that many of Steinberg's results follow from this property alone. We here present the covering theory of perfect groups, which is analogous to the well-known covering theory of nice topological spaces and sheds a new light on Schur's representation groups.

Although the covering theory of perfect groups is quite pleasing, most applications of it require the computation of the Schur multiplicator of interesting groups and this may be quite hard. Such applications are the study of finite simple groups and of algebraic groups and algebraic K-theory. In the first case, most of the interesting information seems to be known, though much of it is still to be published, cf. GRIESS [2]. In the other areas, we can barely mention some of the existing work and research is going on.

5.1 DEFINITION. A group G is called perfect if $G_{ab} = 0$ or, equivalently, $G = [G,G]$. An epimorphism $\pi: G \to Q$ is called a perfect cover of Q, if G (hence Q) is perfect and $Ker(\pi) \subseteq Z(G)$.

Due to the exact sequence I.(3.3'), π is a perfect cover of the perfect group Q precisely when $Ker\ \pi \hookrightarrow G \xrightarrow{\pi} Q$ is a stem extension. In this section, we feel free to identify π with the latter extension.

5.2 EXAMPLES of perfect groups. (a) The most familiar perfect groups are various families of matrix groups, e.g. SL(n,K) and

PSL(n,K) for $n \geq 3$, or for $n = 2$ when $|K| > 3$, cf. HUPPERT [1;II§6]. Here we allow finite as well as infinite fields K, of course.

(b) Every simple non-abelian group is clearly perfect. The smallest non-trivial finite perfect group must be simple, hence is the alternating group $A_5 \cong PSL(2,5)$.

(c) Another family of perfect groups consists of the fundamental groups of homology spheres, i.e. (nice) topological spaces X with integral homology isomorphic to that of a sphere S^n for some $n \geq 2$. Indeed $\pi_1(X)_{ab} \cong H_1(X) \cong H_1(S^n) = 0$. Actually, the result of KAN/THURSTON [1] on the "reverse plus-construction" indicates that perfect groups abound in topology, cf. also BAUMSLAG/DYER/ HELLER [1].

5.3 THEOREM (KERVAIRE [1;§1]). Let Q be a perfect group. Then there is a unique representation group

$$e = e/0 : M(Q) \rightarrowtail^{\varkappa} Q_0 \xrightarrow{\pi} \twoheadrightarrow Q$$

with $\theta_*(e/0) = 1$. For all subgroups U of M(Q) , the extension e/U is a perfect cover of Q. Conversely, every perfect cover $e_1 : N \rightarrowtail G \twoheadrightarrow Q$ of Q is isomorphic over Q to e/U with $U = \text{Ker}\ \theta_*(e_1) \subseteq M(Q)$. Moreover, $\text{Ker}\ \theta_*(e/U) = U$ and $e/\text{Ker}\ \theta_*(e_1)$ is isomorphic to e_1 over Q.

This theorem gives a one-to-one correspondence between the isomorphism classes of perfect covers of Q and the subgroups of M(Q) . In the case of finite perfect Q, the uniqueness of the representation groups is due to SCHUR [1;Satz IV,p.38]. Recall that e/U is defined as an induced extension by 3.1,

$$(5.1) \quad e/U : \frac{M(Q)}{U} \rightarrowtail^{\varkappa_U} Q_U \xrightarrow{\pi_U} \twoheadrightarrow Q\ , \quad Q_U = \frac{Q_0}{U}\ .$$

PROOF (ECKMANN/HILTON/STAMMBACH [1;Thm.5.3]). The uniqueness of the representation group follows from $\mathrm{Ext}(Q_{ab},M(Q)) = 0$ by Proposition 3.2. As $\theta_*(e/0) = 1$, we have $\theta_*(e/U)=\mathrm{nat}\colon M(Q) \to M(Q)/U$ an epimorphism; hence e/U is a perfect cover of Q. Conversely, let $e_1\colon N \rightarrowtail G \twoheadrightarrow Q$ be a perfect cover. By Remark 3.4 (iii) and the uniqueness of the representation group, there exists $U \subseteq M(Q)$ and $\varphi\colon M(Q)/U \cong N$ such that $[e_1] = \varphi_*[e/U]$. In view of $\theta_*(e) = 1$, this implies $\mathrm{Ker}\ \theta_*(e_1) = U$. The reader easily verifies the last assertion. □

5.4 REMARK. Let $\bar{e}\colon R \hookrightarrow F \overset{\rho}{\twoheadrightarrow} Q$ be a free presentation of the perfect group Q. Then the following extension e_o , to be regarded as a "subextension" of $c(\bar{e})$, is congruent to $e/0$,

$$(5.2) \qquad e_o : \frac{R\cap[F,F]}{[R,F]} \hookrightarrow \frac{[F,F]}{[R,F]} \twoheadrightarrow Q .$$

Indeed, e_o is an extension due to $\rho[F,F] = Q$ and clearly central. Let $i\colon M(Q)_{\bar{e}} \to R/[R,F]$ denote the inclusion. The obvious morphism $(i,\cdot,1)\colon e_o \to c(\bar{e})$ of extensions yields $\theta_* c(\bar{e}) = i\cdot\theta_*(e_o)$. Since $\theta_* c(\bar{e}) = \theta_*(\bar{e}) = i$ by Proposition I.3.5 , we conclude $\theta_*(e_o) = 1$. In this context, the representation group $e/0$ resp. e_o is called the <u>universal perfect cover</u> of Q in view of Theorem 5.6 and Prop. 5.10 (iii) below.

5.5 THEOREM (ECKMANN/HILTON/STAMMBACH [1;Thm.5.7]). Let $e = (\varkappa,\pi)\colon N \rightarrowtail G \twoheadrightarrow Q$ be a central extension and X a perfect group. A homomorphism $\gamma\colon X \to Q$ can be lifted to a homomorphism $\beta\colon X \to G$ with $\pi\circ\beta = \gamma$ if, and only if, $\mathrm{Im}\ M(\gamma) \subseteq \mathrm{Im}\ M(\pi)$ in $M(Q)$. The lifting β is unique. If γ is surjective and e is stem, then β is also surjective.

PROOF. By Proposition I.1.12 a lifting exists exactly if

$\gamma^*[e] \in \mathrm{Cext}(X,N)$ is zero. By Proposition I.3.5 and Theorem I.3.8 we obtain:

$$\gamma^*[e] = 0 \Longleftrightarrow \theta_*(e\gamma) = \theta_*(e)\circ M(\gamma) = 0$$
$$\Longleftrightarrow \mathrm{Im}\, M(\gamma) \subseteq \mathrm{Ker}\, \theta_*(e) = \mathrm{Im}\, M(\pi) \;;$$

here θ_* is an isomorphism due to $X_{ab} = 0$. This condition also gives the uniqueness by Theorem I.1.10, as $\mathrm{Der}(X,N,0) = \mathrm{Hom}(X_{ab},N) = 0$. Finally, assume that γ is surjective and e is stem. From $G = \kappa N\cdot(\mathrm{Im}\,\beta)$ and $\kappa N \subseteq Z(G)\cap[G,G]$ we conclude $\kappa N \subseteq [\mathrm{Im}\,\beta, \mathrm{Im}\,\beta]$, hence $\mathrm{Im}\,\beta = G$. □

5.6 THEOREM (KERVAIRE[1;§1]). Let Q be a perfect group, assume the notation of 5.3. For subgroups U and V of $M(Q)$, there exists a homomorphism $\beta\colon Q_U \to Q_V$ with $\pi_V\circ\beta = \pi_U$ precisely when $U \subseteq V$. If so, β is uniquely determined and is a perfect cover of Q_V.

Moreover, $M(\pi_U)\colon M(Q_U) \to M(Q)$ is monic with image U and $\mathrm{Ker}\,\pi_U \cong M(Q)/U$, thus $M(Q_U) \cong U$. Conversely if the group G in a perfect cover $\pi\colon G \to Q$ satisfies $M(G) = 0$, then π is isomorphic over Q to the universal perfect cover.

This theorem implies that the group Q_0 in the universal perfect cover $e/0$ may also serve as "the universal covering group" $(Q_U)_0$ of all the groups Q_U. The key point is $M(Q_0) = 0$. In the case of finite perfect groups Q, the vanishing of $M(Q_0)$ was already shown by IWAHORI/MATSUMOTO [1;§2].

PROOF. Exact sequence (I.3.3') gives $\mathrm{Im}\, M(\pi_U) = \mathrm{Ker}\,\theta_*(e/U) = U$ and $\mathrm{coker}\, M(\pi_U) \cong \mathrm{Ker}\,\pi_U$. Thus β with $\pi_V\circ\beta = \pi_U$ exists, by Theorem 5.5, precisely when $U \subseteq \mathrm{Im}\, M(\pi_V) = V$; such β is unique and surjective. Since $\mathrm{Ker}\,\beta \subseteq \mathrm{Ker}(\pi_U) \subseteq Z(Q_U)$, β is indeed a perfect cover of Q_V. Next, $M(\pi_U)$ is monic by Thm. I.4.4 (ii). We obtain perfect covers $\beta_U\colon Q_0 \to Q_U$ since $0 \subseteq U$ for all subgroups $U \subseteq M(Q)$. Conversely, assume $M(G) = 0$ for the perfect cover $\pi: G \to Q$. Then π is isomorphic over Q to some π_U by Theorem 5.3; and $U \cong M(Q_U) \cong M(G) = 0$ by the above. □

5.7 PROPOSITION. Let $\pi: G \to Q$ be a perfect cover of the perfect group Q, then $\pi Z(G) = Z(Q)$. Consequently, the composition of perfect covers is a perfect cover and $G/Z(G)$ is centerless for every perfect group G.

PROOF. Trivially $\pi Z(G) \subseteq Z(Q)$. Let $z \in Z(Q)$ and $t \in G$ with $\pi(t) = z$. We claim that $[t,g] = 1$ for all $g \in G$ so that t is central. Certainly $[t,g] \in \operatorname{Ker} \pi \subseteq Z(G)$. As it was first observed by GRÜN [1;p.3], the map $g \longmapsto [t,g]: G \to G$ is a homomorphism by Lemma I.4.1. Now

$$[t,[g,h]] = [t,ghg^{-1}h^{-1}] = [t,g]\cdot[t,h]\cdot[t,g]\cdot[t,h]^{-1} = 1 .$$

This suffices since the perfect group G is generated by commutators. Concerning a composition γ of perfect covers, the first assertion implies that $\operatorname{Ker} \gamma$ is again central. □

The preceding results indicate a strong analogy with the covering theory of connected topological spaces, the role of the fundamental group there being played by the Schur multiplicator here. An additional feature is the existence of a "smallest group". In detail, a perfect group G determines a family of perfect groups with two distinguished members. The family consists of the perfect covers G_1 of $Q = G/Z(G)$ or, equivalently, the groups G_0/U where G_0 is the universal covering group and $U \subseteq Z(G_0)$. On one side, G_0 is characterized by $M(G_0) = 0$ and covers all G_1 . On the other extreme, we have $Q \cong G_1/Z(G_1)$ for all G_1 , and Q is characterized by $Z(Q) = 0$.

5.8 EXAMPLE. We determine the perfect covers of A_5. It is well-known that A_5 is isomorphic to the icosahedral group, the latter having a free presentation $\langle x,y : x^3 = y^5 = (xy)^2 = 1\rangle$. Let the binary icosahedral group be defined by the presentation

(5.3) $\quad G = \text{group} \langle s,t : s^3=t^5=(st)^2 \rangle$.

There is an obvious epimorphism $\pi: G \to A_5$ with $\pi(s) = x$ and $\pi(t) = y$. We claim $\text{Ker}\,\pi \cong \mathbb{Z}/2$ and that π is the universal perfect cover of A_5. Thus $|G| = 120$ and $M(G) = 0$ and $M(A_5) = \mathbb{Z}/2$; the family of perfect groups determined by A_5 consists of G and A_5 alone. One may deduce $SL(2,5) \cong G$.

We proceed to prove the claim. The element $z := s^3$ commutes with the generators and hence is central in G; it obviously generates $\text{Ker}\,\pi$ as a normal subgroup. Thus

$$\text{Ker}\,\pi = \{1,z,z^{-1},z^2,z^{-2},\ldots\}.$$

Abelianizing the presentation (5.3), we easily find $G_{ab} = 0$, thus π is a perfect cover. The Sylow subgroups of A_5 are $\mathbb{Z}/2 \times \mathbb{Z}/2$, $\mathbb{Z}/3$, and $\mathbb{Z}/5$. Since $M(\mathbb{Z}/2 \times \mathbb{Z}/2) \cong \mathbb{Z}/2$ and $M(\text{cyclic}) = 0$, the corestriction argument I.6.9 yields $M(A_5) = 0$ or $M(A_5) \simeq \mathbb{Z}/2$. By Theorem 5.6, we have an exact sequence

$$0 \longrightarrow M(G) \longrightarrow M(A_5) \longrightarrow \text{Ker}\,\pi \longrightarrow 0.$$

We are going to show $z \neq 1$. This implies $\text{Ker}\,\pi \cong M(A_5) \cong \mathbb{Z}/2$ and $M(G) = 0$, hence the claim. To this end, we define a 2-dimensional complex (unitary) representation φ of G by

$$\varphi(s) = \tfrac{1}{2}\begin{pmatrix}1-i, & -1+i\\ 1+i, & 1+i\end{pmatrix} \quad\text{and}\quad \varphi(t) = \tfrac{1}{2}\begin{pmatrix}\tau-\sigma i, & -1\\ 1 & ,\tau+\sigma i\end{pmatrix}$$

with $\sigma := (\sqrt{5}-1)/2$ and $\tau := (\sqrt{5}+1)/2 = \sigma^{-1}$. One may check

$$\varphi(s)^3 = \varphi(t)^5 = [\varphi(s)\cdot\varphi(t)]^2 = \begin{pmatrix}-1 & 0\\ 0 & -1\end{pmatrix}.$$

This shows that φ is indeed a representation of G and gives $z \neq 1$. (Our appeal to this representation is not at all accidental, cf.

Example IV.1.8). It is sometimes useful to know that φ is faithful and z is the only element of order two in G. (Proof: Use the simplicity of A_5 and the order of $\varphi(t)$.)

5.9 REMARK. The Schur multiplicators of all finite simple groups have been found, often by exhibiting a universal perfect cover (representation group). For the results see the summary by GRIESS [2] and follow the references given there. Note that the Schur multiplicator of the Mathieu group M_{22} is $\mathbb{Z}/12$ rather than $\mathbb{Z}/6$, according to MAZET [1]. Relative to the existence question, LEMPKEN [1] computed the multiplicator of Janko's simple group J_4 as trivial; later the existence of J_4 has been asserted by S. NORTON [1]. Let us not forget that already SCHUR [2] determined the representation groups of all PSL(2,q) , de SÉGUIER [1] and SCHUR [3] those of the alternating groups.

The following proposition is often handy when you have a candidate $\pi: G \to Q$ for the universal perfect cover, but do not yet know M(Q) resp. M(G) .

5.10 PROPOSITION (Steinberg and Kervaire). Let $\pi: G \to Q$ be a perfect cover of the perfect group Q. Then the following are equivalent.

(i) π is isomorphic over Q to the universal perfect cover of Q;

(ii) every representation group of G splits (as an extension);

(iii) for every central extension $e: A \rightarrowtail X \twoheadrightarrow Q$ by Q, there exists a (necessarily unique) lifting homomorphism $\eta: G \to X$ with $\xi \circ \eta = \pi$.

PROOF. Note the assumption $G_{ab} = 0$. If $G \cong Q_o$, then $M(G) = 0$ by Theorem 5.6. Thus (ii) holds due to

$\mathrm{Cext}(G,A) \cong \mathrm{Hom}(M(G),A) = 0$ while (iii) is immediate from Theorem 5.5. Conversely, assume (ii) and let e be a representation group of G. Then e is direct split and stem at the same time, thus $M(G) \cong \text{kernel} = 0$; then (i) follows by Theorem 5.6. Finally, assume (iii) and let e be a representation group of Q; then $M(G) \cong \mathrm{Im}\, M(\pi) \subseteq \mathrm{Im}\, M(\xi) = \mathrm{Ker}\, \theta_*(e) = 0$ by 5.5 and 5.6. □

5.11 EXAMPLE: the functor K_2 of algebraic K-theory. Let Λ be an arbitrary ring with 1. Let $E(\Lambda)$ be the subgroup of $GL(\Lambda) = \varinjlim GL(n,\Lambda)$ which is generated by the elementary matrices. Here the typical $n{\times}n$ matrix X of $GL(n,\Lambda)$ is identified with the matrix

$$\left(\begin{array}{c|c} X & 0 \\ \hline 0\ldots0 & 1 \end{array}\right)$$

in $GL(n+1,\Lambda)$. A matrix is called elementary, if it agrees with the identity matrix except for at most one off-diagonal entry $\lambda \in \Lambda$; call this matrix e_{ij}^{λ} if λ appears as the (i,j)-th entry . (Think of elementary row and column operations.) By standard matrix calculus, $E(\Lambda)$ is centerless and each $e_{i,j}^{\lambda}$, $i \neq j$, is a commutator. Let the Steinberg group $St(\Lambda)$ be defined by the following free presentation: generators are the symbols x_{ij}^{λ} for all $\lambda \in \Lambda$ and natural numbers $i \neq j$, the defining relations are

(5.4a) $\quad x_{ij}^{\lambda}\, x_{ij}^{\mu} = x_{ij}^{\lambda+\mu} \qquad i \neq j$

(5.4b) $\quad [x_{ij}^{\lambda}, x_{jl}^{\mu}] = x_{il}^{\lambda\mu} \qquad i \neq j \neq l \neq i$

(5.4c) $\quad [x_{ij}^{\lambda}, x_{kl}^{\mu}] = 1 \qquad i \neq j \neq k \neq l \neq i$.

The point is that the corresponding relations hold in $E(\Lambda)$, hence the assigment $x_{ij}^{\lambda} \longmapsto e_{ij}^{\lambda}$ defines an epimorphism $\phi\colon St(\Lambda) \to E(\Lambda)$, the kernel of which is Milnor's $K_2(\Lambda)$ by definition. Due to the relations (5.4b), $St(\Lambda)$ is perfect. KERVAIRE [1] showed that

$$K_2(\Lambda) \hookrightarrow St(\Lambda) \xrightarrow{\phi} E(\Lambda)$$

is the universal perfect cover of $E(\Lambda)$, consequently $M(St(\Lambda)) = 0$ and $K_2(\Lambda) = Z(St(\Lambda)) \cong M(E(\Lambda))$. For the proof, the appeal to 5.10(ii) is most appropriate; see also MILNOR [1;§5]. When Λ is a finite field, then $K_2(\Lambda) = 0$ and the above actually is a free presentation of $E(\Lambda)$, cf. MILNOR [1;Cor.9.13]. On the other hand, consult MILNOR's book [1] and the survey by DENNIS/STEIN [1] for $K_2(\mathbb{Z}) \cong \mathbb{Z}/2$ and other examples of known $K_2(\Lambda)$.

5.12 EXAMPLES: $SL(n,\mathbb{Z})$. The concept of the Steinberg group can also be defined for the Chevalley groups over commutative rings, including $SL(n,\mathbb{Z})$ and $Sp(2n,\mathbb{Z})$. Here the paper of van der KALLEN/STEIN [1] is the last in a series, leading to complete results whenever the Chevalley groups in question are perfect. In particular, the Schur multiplicators of $SL(3,\mathbb{Z})$ and $SL(4,\mathbb{Z})$ are isomorphic to $\mathbb{Z}/2 \times \mathbb{Z}/2$ by van der KALLEN [1]. Combining the results of J.R. Silvester, as presented by MILNOR [1;§10], and of KERVAIRE [1], we find $M(SL(n,\mathbb{Z})) \cong \mathbb{Z}/2$ for $n \geq 5$. Finally $M(SL(2,\mathbb{Z})) = 0$, this will be shown in IV.1.7(d).

5.13 PROPOSITION. For every perfect group G, let

$$e(G)_o : M(G) \rightarrowtail G_o \xrightarrow{\pi_o} G$$

be a fixed universal perfect cover with $\theta_*(e(G)_o) = 1_{M(G)}$.

(a) Every homomorphism $\gamma: G \to H$ between perfect groups yields a lifting $\tilde{\gamma}: G_o \to H_o$ which is unique by $\pi_o\tilde{\gamma} = \gamma\pi_o$. The induced map on the kernels is $\alpha = M(\gamma)$.

$$\begin{array}{cccccc} e(G)_o : & M(G) & \rightarrowtail & G_o & \twoheadrightarrow & G \\ & \downarrow{\scriptstyle\alpha} & & \downarrow{\scriptstyle\tilde{\gamma}} & & \downarrow{\scriptstyle\gamma} \\ e(H)_o : & M(H) & \rightarrowtail & H_o & \twoheadrightarrow & H \end{array}$$

(b) If $e=(\varkappa,\pi)\colon N \rightarrowtail G \twoheadrightarrow Q$ is an extension of perfect groups, then the following sequences are exact:

(5.5a) $$N_o \xrightarrow{\tilde{\varkappa}} G_o \xrightarrow{\tilde{\pi}} Q_o \longrightarrow 0$$

(5.5b) $$M(N) \xrightarrow{M(\varkappa)} M(G) \xrightarrow{M(\pi)} M(Q) \longrightarrow 0 \ .$$

This proposition is due to KERVAIRE [1;§2], our proof follows a suggestion of ECKMANN/HILTON/STAMMBACH [1;p.119]. The uniqueness assertion implies functoriality. Thus, whenever α is isomorphic, so is $\tilde{\alpha}$.

PROOF. (a) As $M(G_o) = 0$ by 5.6, the first assertion is immediate from 5.5. The naturality of θ_* gives $\alpha = M(\gamma)$.

(b) As in I.(3.4), let $\rho\colon F \twoheadrightarrow G$ be a free presentation of G with $\operatorname{Ker} \rho = S$ and $R = \operatorname{Ker}(\pi\rho)$. Then $S \hookrightarrow R \twoheadrightarrow N$ is a free presentation of N. Without loss of generality, we choose $e(G)_o$ etc. as specified in (5.2). Then

$$G_o = \frac{[F,F]}{[F,S]} \ , \quad Q_o = \frac{[F,F]}{[F,R]} \ , \quad N_o = \frac{[R,R]}{[R,S]} \ ;$$

the unique maps $\tilde{\pi}$ and $\tilde{\varkappa}$ are induced in the obvious way. Clearly $\tilde{\pi}$ is surjective with kernel $[F,R]/[F,S]$. The assertion $\operatorname{Ker} \tilde{\pi} = \operatorname{Im} \tilde{\varkappa}$ is equivalent to $[F,R] = [R,R]\cdot[F,S]$. Since N and G are perfect, we have $R = S\cdot[R,R]$ and $F = S\cdot[F,F]$. As the groups R etc. are normal in F, we conclude

$$[F,R] \subseteq [R,S]\cdot[[F,F],S]\cdot[[F,F],[R,R]] \subseteq [R,R]\cdot[F,S] \ .$$

The other inclusion is trivial. Finally, the Ker-Coker lemma for

$$\begin{array}{ccccccccc}
 & & N_o & \longrightarrow & G_o & \longrightarrow & Q_o & \longrightarrow & 0 \\
 & & {\scriptstyle \pi_o}\downarrow & & \downarrow & & \downarrow & & \\
0 & \longrightarrow & N & \xrightarrow{\varkappa} & G & \xrightarrow{\pi} & Q & \longrightarrow & 0
\end{array}$$

gives the exactness of (5.5b). □

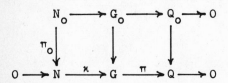

CHAPTER III. ISOCLINISM

1. Isoclinic Groups and Central Extensions

Throughout this section we denote by

$$e : N \overset{\varkappa}{\rightarrowtail} G \overset{\pi}{\twoheadrightarrow} Q$$

$$e_i : N_i \overset{\varkappa_i}{\rightarrowtail} G_i \overset{\pi_i}{\twoheadrightarrow} Q_i \quad , \ i=1,2$$

central extensions. Let us denote by c the commutator function of e, i.e.

$$c : Q \times Q \to [G,G] \ ; \quad (\pi g, \pi h) \longmapsto [g,h] \ ,$$

and by c_i ,i=1,2 the commutator function of e_i ; if Q is abelian, then $[G,G] \subseteq \varkappa N$, and the map $\varkappa^{-1}c$ coincides with the commutator form of I.3.10.

1.1 DEFINITION. The central extensions e_1 and e_2 are called <u>isoclinic</u>, precisely when there exist isomorphisms $\eta: Q_1 \rightarrowtail\!\!\!\twoheadrightarrow Q_2$ and $\xi: [G_1,G_2] \rightarrowtail\!\!\!\twoheadrightarrow [G_2,G_2]$, such that the following diagram is commutative:

$$\begin{array}{ccc} Q_1 \times Q_1 & \xrightarrow{c_1} & [G_1,G_1] \\ \downarrow{\scriptstyle \eta\times\eta} & & \downarrow{\scriptstyle \xi} \\ Q_2 \times Q_2 & \xrightarrow{c_2} & [G_2,G_2] \end{array} ,$$

i.e. $\xi[g_1,h_1] = [g_2,h_2]$ for all $g_1,h_1 \in G_1$, $\pi_2 g_2 = \eta\pi_1 g_1$, $\pi_2 g_2 = \eta\pi_1 h_2$. The pair (η,ξ) is called an <u>isoclinism</u> from e_1 to e_2 , denoted by $(\eta,\xi): e_1 \sim e_2$.

1.2 REMARK. In 1.1 it suffices to assume that ξ is a monomorphism. Furthermore, η determines ξ uniquely.

Let G be a group. Then we obtain the following central extension:

(1.1) $e_G : Z(G) \hookrightarrow G \xrightarrow{nat} G/Z(G)$.

As mentioned above, isoclinism of groups was defined in HALL [1]. In terms of 1.1, it reads as follows:

1.3 DEFINITION. Let G and H be groups. Then G and H are called <u>isoclinic</u>, precisely when e_G and e_H are isoclinic. An isoclinism (η, ξ) from e_G to e_H is also called an <u>isoclinism</u> from G to H, notation: $(\eta, \xi): G \sim H$.

1.4 PROPOSITION. Let $(\eta, \xi): e_1 \sim e_2$ be an isoclinism.

(i) η induces an isomorphism $\eta': G_1/Z(G_1) \rightarrowtail\!\!\!\!\twoheadrightarrow G_2/Z(G_2)$, and (η', ξ) is an isoclinism from G_1 to G_2 .

(ii) $\varkappa_1(N_1) = Z(G_1)$, if and only if $\varkappa_2(N_2) = Z(G_2)$.

PROOF. Let $g_1 \in Z(G_1)$, then $[g_1, g] = 1$ for all $g \in G$. By 1.1 this holds, if and only if $[g_2, h] = 1$ for all $h \in G_2$ and g_2 with $\pi_2 g_2 = \eta\pi_1 g_1$. This proves that η maps $\pi_1 Z(G_1)$ onto $\pi_2 Z(G_2)$. Now, (i) and (ii) follow immediately. □

1.5 REMARKS. One readily sees that the composition of isoclinisms is again an isoclinism, and that $(\eta, \xi): e_1 \sim e_2$ implies $(\eta^{-1}, \xi^{-1}): e_2 \sim e_1$. Thus isoclinism is an equivalence relation.

Isoclinisms of the form $(\eta, \xi): e \sim e$, resp. $(\eta, \xi): G \sim G$ are called <u>autoclinisms</u> of e resp. G, and they constitute a group Acl(e) , resp. Acl(G). Each isoclinism from e_1 to e_2 induces an isomorphism from $Acl(e_1)$ to $Acl(e_2)$.

Let $(\alpha, \beta, \gamma): e_1 \rightarrowtail\!\!\!\!\twoheadrightarrow e_2$ be an isomorphism and β' the restriction of β to $[G_1, G_1]$. Then (γ, β') is an isoclinism from e_1 to e_2 .

The equivalence classes under isoclinism are called "isoclinism classes", the classes of isoclinic groups were called "families" by P. Hall. By 1.4 the families can be regarded as those isoclinism classes that consist of central extensions e with $\varkappa(N) = Z(G)$, i.e. e is isomorphic to e_G .

The groups which are isoclinic to the trivial group are exactly the abelian groups, which thus form a family. An easy example for two non-abelian isoclinic groups is given by the dihedral group D_8 and the quaternion group Q_8.

Let $(\eta,\xi)\colon e_1 \sim e_2$ be an isoclinism. Then the backward induced extension

$$\eta^*e_2 : N_2 \xrightarrow{\varkappa_2} G_2 \xrightarrow{\eta^{-1}\pi_2} Q_1$$

is isomorphic to e_2 , and $(1_{Q_1},\xi)\colon e_1 \sim \eta^*e_2$ is an isoclinism. Thus in many cases we can restrict ourselves to isoclinisms (η,ξ) in which η is an identity map.

1.6 PROPOSITION. Let $(\eta,\xi)\colon e_1 \sim e_2$ be an isoclinism. Then

(i) $\pi_2\xi g' = \eta\pi_1 g'$, $g' \in [G_1,G_1]$,

(ii) $\xi(\varkappa_1 N_1 \cap [G_1,G_1]) = \varkappa_2 N_2 \cap [G_2,G_2]$.

(iii) $\xi(gxg^{-1}) = h\xi(x)h^{-1}$, $x \in [G_1,G_1]$, $y \in G$, $h \in H$ with $\eta\pi_1 g = \pi_2 h$; i.e. $\xi\colon [G_1,G_1] \rightarrowtail\!\!\!\twoheadrightarrow [G_2,G_2]$ is an operator isomorphism.

PROOF. (i) By 1.1, the assertion holds for commutators, and hence for all elements of $[G_1,G_1]$.

(ii) Let $g' \in (\varkappa_1 N_1 \cap [G_1,G_1])$. Then (i) yields $1 = \eta\pi_1 g' = \pi_2\xi g'$, which implies $\xi g' \in (\varkappa_2 N_2 \cap [G_2,G_2])$. The converse follows analogously.

(iii) It is sufficient to show that the assertion holds for commutators $x = [g_1,g_2]$. Let $h_i \in G_2$ with $\pi_2 h_i = \eta\pi_1 g_i$. Then

$\pi_2(hh_ih^{-1}) = \eta\pi_2(gg_ig^{-1})$ and therefore

$$\xi(gxg^{-1}) = \xi([gg_1g^{-1},gg_2g^{-1}]) = [hh_1h^{-1},hh_2h^{-1}] =$$
$$= h[h_1,h_2]h^{-1} = hxh^{-1} . \quad \square$$

As we have mentioned in 1.5, isomorphisms of central extensions induce isoclinisms. This observation gives rise to the following:

1.7 DEFINITION. A morphism (α,β,γ): $e_1 \to e_2$ is called <u>isoclinic</u>, if there exists an isomorphism β': $[G_1,G_1] \rightarrowtail\!\!\twoheadrightarrow [G_2,G_2]$ with (γ,β'): $e_1 \sim e_2$. If β is in addition an epi- resp. monomorphism, (α,β,γ) is called an <u>isoclinic</u> <u>epi</u>- resp. <u>monomorphism</u>.

1.8 PROPOSITION. Let (α,β,γ): $e_1 \to e_2$ be a morphism.

(i) (α,β,γ) is isoclinic, precisely when γ is isomorphic and Ker $\beta \cap [G_1,G_1] = 0$.

(ii) If (α,β,γ) is isoclinic and β' as in 1.7, then $\beta' = \beta|_{[G_1,G_1]}$.

PROOF. As (α,β,γ) is a morphism of extensions, we have $\beta[g_1,g_2] = [h_1,h_2]$, where $g_i \in G_1$, $h_i \in G_2$, $\gamma\pi_1g_i = \pi_2h_i$. This proves (ii). Hence we obtain that (α,β,γ) is isoclinic, if and only if γ is an isomorphism and $\beta|_{[G_1,G_1]}$: $[G_1,G_1] \to [G_2,G_2]$ is a monomorphism, cf. 1.2. The last condition is equivalent to Ker $\beta \cap [G_1,G_1] = 0$. $\square$

1.9 PROPOSITION. Let G and H be groups and β: G $\to$ H a homomorphism. Then β induces an isoclinic morphism from e_G to e_H , if and only if Ker $\beta \cap [G,G] = 0$ and $\mathrm{Im}(\beta)Z(H) = H_i$; in this case we call β an <u>isoclinic homomorphism</u>.

PROOF. The only-if-part is obvious by 1.8. Assume Ker $\beta \cap [G,G] = 0$ and $\mathrm{Im}(\beta)Z(H) = H$. Let $g \in G$ with

$\beta(g) \in Z(H)$. Hence we have $\beta[g,g'] = 1$ for all $g' \in G$, and $\operatorname{Ker}\beta \cap [G,G] = 0$ implies $[g,g'] = 1$, proving $g \in Z(G)$. If $g \in Z(G)$, we have $\beta g \in Z(\operatorname{Im}\beta)$. As $\operatorname{Im}(\beta)Z(H) = H$, we obtain $Z(\operatorname{Im}(\beta)) = Z(H) \cap \operatorname{Im}(\beta)$, and thus $\beta g \in Z(H)$. Now we have shown that β induces a monomorphism from $G/Z(G)$ into $H/Z(H)$, while $\operatorname{Im}(\beta)Z(H) = H$ implies that it is an isomorphism, and we are done by 1.8. □

1.10 REMARKS. (i) A morphism $(\alpha,\beta,\gamma): e_1 \to e_2$ is isoclinic, if and only if γ is an isomorphism and $\beta: G_1 \to G_2$ is an isoclinic homomorphism of groups.

(ii) The composition of isoclinic morphisms is an isoclinic morphism.

(iii) Each isoclinic morphism is a composition of an isoclinic epimorphism and an isoclinic monomorphism.

In the following we consider an isomorphism $\eta: Q_1 \to Q_2$ and put $Q = Q_1$. Then we have the following central extension

$$\tilde{e} = (e_1 \times \eta^* e_2)\Delta_Q : N_1 \times N_2 \rightarrowtail^{\lambda} \tilde{G} \twoheadrightarrow^{\rho} Q ,$$

with the fibre product $\tilde{G} = \{ (g_1,g_2) \mid g_i \in G_i,\ \eta\pi_1 g_1 = \pi_2 g_2 \}$, sometimes denoted by $G_1 \curlywedge G_2$, and $\lambda = \varkappa_1 \times \varkappa_2$, $\rho = \{ (g_1,g_2) \longmapsto \pi_1 g_1 \}$, cf. I.(2.2). Furthermore we denote by $\sigma_i: N_1 \times N_2 \twoheadrightarrow N_i$ and $\tau_i: \tilde{G} \twoheadrightarrow G_i$ the i-th projections and put $\gamma_1 = 1_Q : Q \rightarrowtail\!\!\!\twoheadrightarrow Q_1$, $\gamma_2 = \eta: Q \rightarrowtail\!\!\!\twoheadrightarrow Q_2$. Having fixed the notation, we can readily prove the following proposition, cf. JONES/WIEGOLD [2].

1.11 PROPOSITION. (i) The diagrams

$$(1.2) \qquad \begin{array}{lccccc} \tilde{e} : & N_1 \times N_2 & \xrightarrow{\lambda} & \tilde{G} & \xrightarrow{\rho} & Q \\ & \downarrow \sigma_i & & \downarrow \tau_i & & \downarrow \gamma_i \\ e_i : & N_i & \xrightarrow{\varkappa_i} & G_i & \xrightarrow{\pi_i} & Q_i \end{array}$$

are commutative.

(ii) The isomorphism $\eta \colon Q_1 \rightarrowtail\!\!\!\twoheadrightarrow Q_2$ induces an isoclinism from e_1 to e_2, if and only if $(\sigma_i, \tau_i, \gamma_i)$, $i=1,2$ are isoclinic epimorphisms.

PROOF. The proof of (i) is trivial. Assume that η induces an isoclinism from e_1 to e_2. Then we have an isomorphism $\xi \colon [G_1,G_1] \to [G_2,G_2]$ with $\xi[g_1,g_2] = [h_1,h_2]$, $g_i \in G_1$, $h_i \in G$, $\eta\pi_1 g_i = \pi_2 h_i$. Thus, $(g_i,h_i) \in \tilde{G}$, and we obtain

$$[(g_1,h_1),(g_2,h_2)] = ([g_1,g_2],\xi[g_1,g_2]) .$$

This implies

$$[\tilde{G},\tilde{G}] = \{ (g',\xi g') \mid g' \in [G_1,G_1] \} .$$

As $\operatorname{Ker} \tau_1 = \lambda N_2$ and $\operatorname{Ker} \tau_2 = \lambda N_1$, we obtain $\operatorname{Ker} \tau_i \cap [\tilde{G},\tilde{G}] = 0$. By 1.8, $(\sigma_i, \tau_i, \gamma_i)$ are isoclinic epimorphisms. As $\eta = \gamma_2\gamma_1^{-1}$, the converse follows from Definition 1.7. □

Let $e \colon N \overset{\varkappa}{\rightarrowtail} G \overset{\pi}{\twoheadrightarrow} Q$ be a central extension and A an abelian group. We denote by φ the projection of $G \times A$ onto G and we obtain the central extension

$$e \times A : N \times A \overset{\varkappa\times 1}{\rightarrowtail} G \times A \overset{\pi\varphi}{\twoheadrightarrow} Q .$$

Let $\mu = \{g \longmapsto (g,1)\} \colon G \to G{\times}A$, define $\varphi' \colon N \times A \twoheadrightarrow N$ and $\mu' \colon N \rightarrowtail N \times A$ similarly. Then the diagrams

$$\begin{array}{cccccc}
e \times A : & N \times A & \overset{\varkappa\times 1}{\rightarrowtail} & G \times A & \overset{\pi\varphi}{\twoheadrightarrow} & Q \\
& \varphi' \downarrow \uparrow \mu' & & \varphi \downarrow \uparrow \mu & & \| \\
e \quad : & N & \overset{\varkappa}{\longrightarrow} & G & \twoheadrightarrow & Q
\end{array}$$

are commutative, $(\varphi',\varphi,1_Q) \colon e{\times}A \to e$ is an isoclinic epimorphism, and $(\mu',\mu,1_Q) \colon e \to e{\times}A$ is an isoclinic monomorphism.

Let $e_1, e_2, \tilde{e}, \sigma_i, \tau_i, \gamma_i$ be as above, $\eta \colon Q_1 \rightarrowtail\!\!\!\twoheadrightarrow Q_2$ an isomorphism, and $A = \tilde{G}_{ab}$, and

$$\alpha = \{ n \longmapsto (\sigma_1 n, ab(\lambda n)) \} : N_1 \times N_2 \longrightarrow N_1 \times A$$

$$\beta = \{ g \longmapsto (\tau_1 g, ab(g)) \} : \tilde{G} \longrightarrow G_1 \times A .$$

1.12 LEMMA. Assume that η induces an isoclinism from e_1 to e_2. Then (α,β,γ_1): $\tilde{e} \to e_1 \times A$ is an isoclinic monomorphism.

PROOF. The commutativity of (1.2) implies that (α,β,γ_1) is a morphism. Let $g = (g_1,g_2) \in \operatorname{Ker} \beta$, and denote by ξ the isomorphism from $[G_1,G_1]$ to $[G_2,G_2]$ induced by η, i.e. (η,ξ): $e_1 \sim e_2$. Then $\tau_1(g) = g_1 = 1$ and $ab(g) = 1$, i.e. $(g_1,g_2) \in [\tilde{G},\tilde{G}]$. From the proof of 1.11 we obtain $[G,G] = \{ (g_1, \xi g_1) \mid g_1 \in [\tilde{G}_1,\tilde{G}_1] \}$, and it follows $g_2 = 1$. Thus we have $\operatorname{Ker} \beta = 0$, and by 1.8 we obtain that (α,β,γ_1) is an isoclinic monomorphism. □

Let (η,ξ): $e_1 \sim e_2$ be an isoclinism. Now we consider the commutative diagram (1.3) below , which will yield two characterizations of isoclinism. The following holds:

$$\alpha N_1 = \{ (n_1, ab(\varkappa_1 n_1, 1) \mid n_1 \in N_1 \} ,$$

$$\operatorname{Ker}(nat) = \{ (\varkappa_1 n_1, ab(\varkappa_1 n_1, 1)) \mid n_1 \in N_1 \} ,$$

$$[\tilde{G},\tilde{G}] = \{ (g, \xi g) \mid g \in [G_1,G_1] \} .$$

This yields

$$\operatorname{Ker}(nat) \cap [G_1 \times A, G_1 \times A] = 0 ,$$

$$\alpha N_1 \cap \mu' N_1 = 0 .$$

Furthermore we have $\operatorname{Ker}(nat'\alpha) = N_1 = \operatorname{Ker} \sigma_2$, and we obtain:

1.13 LEMMA. (i) The composition of $(nat', nat, 1_{Q_1})$ and $(\mu', \mu, 1_{Q_1})$ is an isoclinic monomorphism from e_1 into $nat'(e_1 \times A)$.

(ii) The composition of $(nat', nat, 1_{Q_1})$ and (α,β,γ_1) induces

Diagram (1.3)

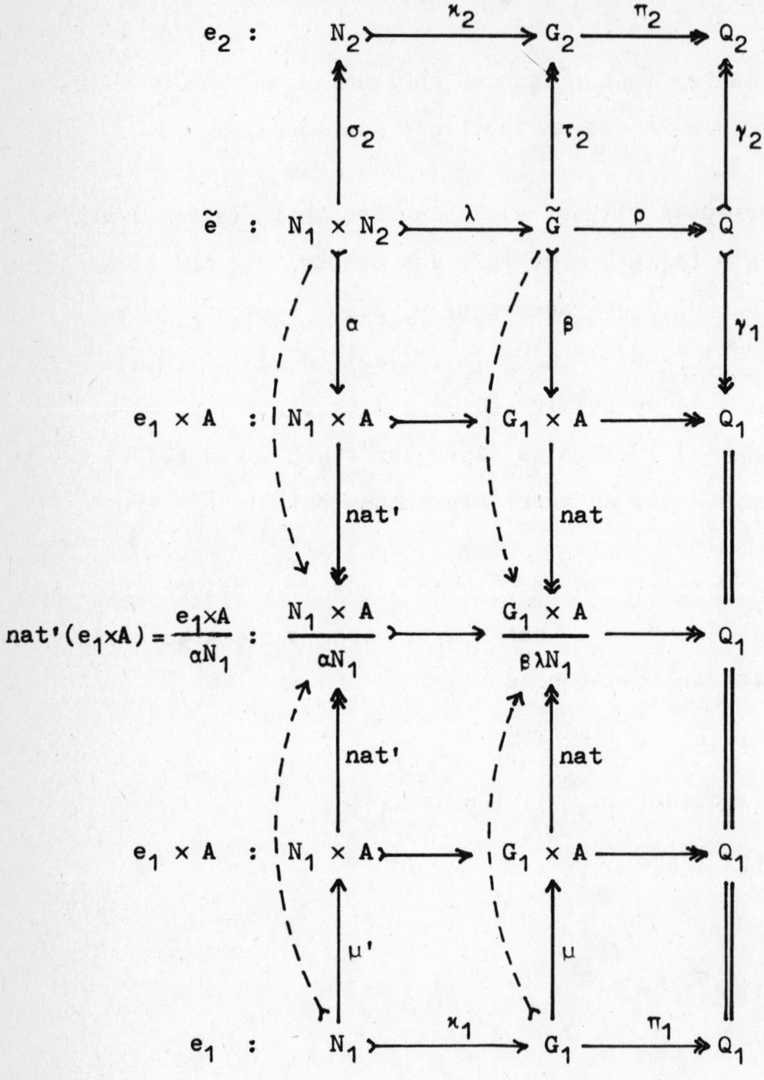
e₂ :
N₂
κ₂
G₂
π₂
Q₂
σ₂
τ₂
γ₂
ẽ :
N₁ × N₂
λ
G̃
ρ
Q̂
α
β
γ₁
e₁ × A :
N₁ × A
G₁ × A
Q₁
nat'
nat
nat'(e₁×A) = e₁×A / αN₁ :
N₁ × A / αN₁
G₁ × A / βλN₁
Q₁
nat'
nat
e₁ × A :
N₁ × A
G₁ × A
Q₁
μ'
μ
e₁ :
N₁
κ₁
G₁
π₁
Q₁

an isoclinic monomorphism from e_2 into $\mathrm{nat}'(e_1 \times A)$. □

The following theorems summarize the results above, the proofs can be easily read from diagram (1.3).

1.14 THEOREM. The following properties are equivalent:

(i) The central extensions e_1 and e_2 are isoclinic.

(ii) There exists a central extension e' together with isoclinic epimorphisms from e' onto e_1 and e_2.

(iii) There exists a central extension e'' together with isoclinic monomorphisms from e_1 and e_2 into e''. □

1.15 REMARK. If e_1 and e_2 are finite, extensions e' and e'' in 1.14 can also be chosen finite. The situations in 1.14 can be roughly sketched by $e_1 \twoheadleftarrow e' \twoheadrightarrow e_2$ and $e_1 \rightarrowtail e'' \leftarrowtail e_2$. In particular, the equivalence relations for central extensions generated by isoclinic epimorphism, resp. isoclinic monomorphism coincide with isoclinism.

1.16 THEOREM. The following properties are equivalent:

(i) e_1 and e_2 are isoclinic.

(ii) There exists an abelian group A, a central extension e', an isoclinic monomorphism from e' into $e_1 \times A$, and an isoclinic epimorphism from e' onto e_2.

(iii) There exists an abelian group B, a central extension e'', an isoclinic epimorphism from $e_1 \times B$ onto e'', and an isoclinic monomorphism from e_2 into e''. □

The statements of 1.16 can be sketched as follows:

$$e_1 \rightleftarrows e_1 \times A \leftarrowtail e' \twoheadrightarrow e_2$$

$$e_1 \rightleftarrows e_1 \times B \twoheadrightarrow e'' \leftarrowtail e_2 \; .$$

For isoclinic groups we obtain:

1.17 PROPOSITION (Hall). Let G be a group.

(i) Let A be an abelian group. Then G and $G \times A$ are isoclinic.

(ii) Let N be a normal subgroup of G. The natural homomorphism nat: $G \twoheadrightarrow G/N$ is an isoclinic epimorphism, precisely if $N \cap [G,G] = 0$.

(iii) If $[G,G]$ is finite, then isoclinism of G and G/N is equivalent to $N \cap [G,G] = 0$.

(iv) Let U be a subgroup of G. The embedding of U into G is an isoclinic monomorphism, if and only if $U \cdot Z(G) = G$.

(v) If $G/Z(G)$ is finite, then isoclinism of G and U is equivalent to $U \cdot Z(G) = G$.

Remark: The assertions (iii) and (v) do not hold in general.

PROOF. The proof of (i),(ii),(iv) is obvious.

(iii) let $[G,G]$ be finite and G isoclinic to G/N . Then $|[G,G]| = |[G/N,G/N]| = |[G,G]N/N| = |[G,G]|/|[G,G] \cap N|$, and we obtain $[G,G] \cap N = 0$.

(v) Assume that $G/Z(G)$ is finite and G isoclinic to U. Then $|G/Z(G)| = |U/Z(U)| \leq |U/Z(G) \cap U| = |UZ(G)/Z(G)| \leq |G/Z(G)|$. This implies $U \cdot Z(G) = G$. □

If we restrict 1.16 to isoclinism of groups, we obtain the following theorem, the equivalence of (i) and (ii) of which is due to WEICHSEL [1].

1.18 THEOREM. Let G and H be groups. Then the following properties are equivalent:

(i) G and H are isoclinic.

(ii) There exist an abelian group A, a subgroup U of $G \times A$ with $UZ(G \times A) = G \times A$, and a normal subgroup N of U with $N \cap [U,U] = 0$, such that U/N is isomorphic to H.

(iii) There exist an abelian group B, a normal subgroup M of $G \times B$ with $M \cap [G \times B, G \times B] = 0$, and a subgroup V of $G \times B/M$ with $VZ(G \times B/M) = G \times B/M$, such that V is isomorphic to H. □

1.19 REMARK. Let $\beta: G \twoheadrightarrow H$ be an isoclinic epimorphism. Then β induces an isoclinism $(\eta, \xi): G \sim H$, where $\xi = \beta|_{[G,G]}$. Hence we have the following commutative diagram with exact rows

$$(1.4)\qquad \begin{array}{ccccc} [H,H] & \xrightarrow{\ \xi^{-1}\ } & G & \twoheadrightarrow & A \\ \| & & \downarrow{\scriptstyle \beta} & & \downarrow \\ [H,H] & \hookrightarrow & H & \twoheadrightarrow & H_{ab} \end{array},$$

where $A \simeq G_{ab}$. On the other hand, if we fix H, while A is any abelian group and G a group which fits into (1.4) for some epimorphism β, then $[G,G] = \mathrm{Im}\, \xi^{-1}$ and $\mathrm{Ker}\, \beta \cap [G,G] = 0$. This shows that β is an isoclinic epimorphism. Hence, (1.4) determines all groups which map epi-isoclinic onto H.

In KING [1], groups G without a normal subgroup N satisfying $N \cap [G,G] = 0$ are called quotient irreducible, and groups G without subgroups U such that $U \cdot Z(G) = G$, are called subgroup irreducible. King proves:

1.20 THEOREM. (i) The group G is quotient irreducible, if and only if the socle of G is contained in $[G,G]$ and $Z(G)/(Z(G) \cap [G,G])$ is a torsion group.

(ii) If G is subgroup irreducible, then $Z(G)$ is contained in the Frattini subgroup of G. The converse holds, if $Z(G)/(Z(G) \cap [G,G])$ is finitely generated.

(iii) If $Z(G) \subseteq [G,G]$, then G is quotient and subgroup irreducible. □

For finite groups, irreducibility of G means that G does not have a subgroup or factor groups being isoclinic to G. One might expect that a finite group which is quotient and subgroup irreducible, is a group of minimal order in its family. But the metacyclic groups in the following example show that this conjecture does not hold, cf. KING [1].

1.21 EXAMPLE. Let p be a prime and $n \geq 2$ an integer. Let us define:

$$G_n = \text{group} \langle a,b : a^{p^n}=b^{p^n}=1 , [a,b]=a^{p^{n-1}} \rangle .$$

We can easily verify

$$Z(G_n) = \langle a^p \rangle \cong \mathbb{Z}/p^{n-1} , [G_n,G_n] = \langle a^{p^{n-1}} \rangle \cong \mathbb{Z}/p$$

$$G_n/Z(G_n) \cong \mathbb{Z}/p \times \mathbb{Z}/p , M(G_n/Z(G_n)) \cong \mathbb{Z}/p$$

$$\text{Im } \theta_*(e_{G_n}) = [G_n,G_n] \cap Z(G_n) = \langle a^{p^{n-1}} \rangle \cong \mathbb{Z}/p .$$

Hence, $\theta_*(e_{G_n})$ is a monomorphism, which proves that e_{G_n} is a generalized representation group of $\mathbb{Z}/p \times \mathbb{Z}/p$, cf. II.2.13. In 2.4 it will be shown that all generalized representation groups of a given group are mutually isoclinic, cf. IV.2.15. Using 1.20, it can be verified that G_n is subgroup and quotient irreducible for all $n \geq 2$.

Let us consider a family which contains finite representatives. From 1.6 we obtain that a group G in this family has minimal order, whenever $Z(G) \subseteq [G,G]$ holds, i.e. if e_G is a stem extension.

P. Hall called these groups "stem groups". In 2.6 it will be shown that each family contains stem groups.

Let G be a group and U a subgroup of G with $U\cdot Z(G) = G$. Hence we can choose a transversal $(g_i)_{i\in I}$ to U in G with $g_i \in Z(G)$. We denote by $K_G(g)$, resp. $K_U(u)$ the conjugacy class of g in G, resp. of u in U. For all $u \in U$ and all g_i , we have $K_G(g_iU) = g_iK_U(u)$. Hence each class of U corresponds to $|I|$ classes of G having the same length. Now we assume that G is finite, and let π be a set of primes. It is easy to see that one can choose the system (g_i) such that g_i is a π-element, precisely when $\bar{g}_i := g_iU$ is a π-element of G/U. (Decompose g_i in a unique product of a π-element and a π'-element, which commute.) Hence, g_iu is a π-element, if and only if u and $\bar{g}_i$ are π-elements. Let $k_n^\pi(G)$, resp. $k_n^\pi(U)$ denote the number of conjugacy classes of π-elements in G, resp. U having length n. Then we have

$$k_n^\pi(G) = k_n^\pi(U)\cdot|G:U|_\pi ,$$

where $|G:U|_\pi$ denotes the π-part of $|G:U|$.

1.22 PROPOSITION. Let G and H be finite isoclinic groups, n a positive integer, and π a set of primes. Then $k_n^\pi(G)\cdot|H|_\pi = k_n^\pi(H)\cdot|G|_\pi$.

PROOF. By 1.14 we can regard G and H as subgroups of a group E with $G\cdot Z(E) = E$ and $H\cdot Z(E) = E$. Our considerations above show

$$k_n^\pi(G)\cdot|E:G|_\pi = k_n^\pi(E) = k_n^\pi(H)\cdot|E:H|_\pi .$$

Using $|E:G|_\pi = |E|_\pi/|G|_\pi$ and $|E:H|_\pi = |E|_\pi/|H|_\pi$, the assertion follows. □

Let $k_n(G)$ denote the number of all classes of G having length n. From 1.22 we obtain the following result of P. Hall:

1.23 COROLLARY. Let G and H be finite isoclinic groups. Then $k_n(G)\cdot|H| = k_n(H)\cdot|G|$. $\square$

We close this section with a few remarks on the correspondence given by the isomorphism of the central factor groups of two isoclinic groups. Let $(\eta,\xi)\colon G \sim H$ be an isoclinism. For all subgroups U of G with $U \supseteq Z(G)$ we denote by ηU the subgroup of H, which is defined by $\eta U/Z(H) = \eta(U/Z(G))$. It is easy to see that $(\eta|_{U/Z(G)},\xi)$ is an isoclinism from $Z(G) \hookrightarrow U \twoheadrightarrow U/Z(G)$ to $Z(H) \hookrightarrow \eta U \twoheadrightarrow \eta U/Z(H)$, and it induces an isoclinism from U to ηU . If G and H are finite, P is a Sylow p-subgroup of U, $V := PZ(G)$, and Q a Sylow p-subgroup of ηV , then Q is also a Sylow p-subgroup of ηU . It can also be shown that η induces an isoclinism from P to Q.

Isoclinism has many invariants, in particular solvability and nilpotency. Let

$$0 \subseteq G^1 = Z(G) \subseteq G^2 \subseteq G^3 \subseteq \dots$$

be the upper central series of G, and

$$G_1 = G \supseteq G_2 = [G,G] \supseteq G_3 \supseteq G_4 \supseteq \dots$$

the lower central series of G, and $(\eta,\xi)\colon G \sim H$. Then

$$0 \subseteq \eta G^2 \subseteq \eta G^2 \subseteq \eta G^3 \subseteq \dots$$

is the upper central series of H, and 1.6 yields that

$$H \supseteq \xi[G,G] = [H,H] \supseteq \xi G_3 \supseteq \xi G_4 \supseteq \dots$$

is the lower central series of H.

2. Isoclinism and the Schur Multiplicator

In this section we study the connection between isoclinism and the subgroups of the Schur multiplicators. Assume as in Section 1 that

$$e : N \rightarrowtail^{\varkappa} G \overset{\pi}{\twoheadrightarrow} Q$$

$$e_i : N_i \rightarrowtail^{\varkappa_i} G_i \overset{\pi_i}{\twoheadrightarrow} Q_i \ , \ i=1,2$$

are central extensions. Let us consider the homomorphism $\varkappa\theta_*(e): M(Q) \to G$. We have

$$\mathrm{Im}(\varkappa\theta_*(e)) = \varkappa N \cap [G,G] \ .$$

Thus $\varkappa\theta_*(e)$ induces a homomorphism $\theta': M(Q) \to [G,G]$, such that the sequence

$$(2.1) \qquad \mathrm{Ker}\ \theta_*(e) \hookrightarrow M(Q) \xrightarrow{\theta'} [G,G] \overset{\pi'}{\twoheadrightarrow} [Q,Q]$$

is exact, π' being induced by $\pi: G \twoheadrightarrow Q$. The naturality of $\theta_*(e)$, cf. I.3.5, yields the naturality of (2.1) with respect to morphisms of central extensions. This is quite useful for studying isoclinic morphisms.

2.1 LEMMA. Let $(\alpha,\beta,\gamma): e_1 \to e_2$ be a morphism of central extensions and assume that $\gamma: Q_1 \to Q_2$ is isomorphic. Then the following conditions are equivalent:

(i) (α,β,γ) is isoclinic .

(ii) $M(\gamma)\ \mathrm{Ker}\ \theta_*(e_1) = \mathrm{Ker}\ \theta_*(e_2)$.

PROOF. For any morphism $(\alpha,\beta,\gamma) : e_1 \to e_2$, the naturality of (2.1) yields the commutativity of the following diagram

$$\begin{array}{ccccccc}
\operatorname{Ker}\theta_*(e_1) & \hookrightarrow & M(Q_1) & \xrightarrow{\theta_1'} & [G_1,G_1] & \twoheadrightarrow & [Q_1,Q_1] \\
\downarrow & & \downarrow{\scriptstyle M(\gamma)} & & \downarrow{\scriptstyle \beta'} & & \downarrow{\scriptstyle \gamma'} \\
\operatorname{Ker}\theta_*(e_2) & \hookrightarrow & M(Q_2) & \xrightarrow{\theta_2'} & [G_2,G_2] & \twoheadrightarrow & [Q_2,Q_2]
\end{array},$$

where θ_i' , $i=1,2$ are defined as above , $\beta' = \beta|_{[G_1,G_1]}$, $\gamma' = \gamma|_{[Q_1,Q_1]}$. As γ is isomorphic, the same holds for γ' and $M(\gamma)$, the restriction of $M(\gamma)$ to $\operatorname{Ker}\theta_*(e_1)$ is monomorphic, and β epimorphic. By Proposition 1.8 , condition (i) of this lemma holds, if and only if β' is isomorphic, being equivalent to (ii) by the diagram above. □

2.2 LEMMA. Let $\eta: Q_1 \to Q_2$ be an isomorphism and, as in Section 1, let $\tilde{e} = (e_1 \times \eta^* e_2)\Delta_Q$. Then

$$\operatorname{Ker}\theta_*(\tilde{e}) = M(\eta)^{-1}\operatorname{Ker}\theta_*(e_2) \cap \operatorname{Ker}\theta_*(e_1) .$$

PROOF. From (1.2) and the naturality of θ_* we obtain

$$\theta_*(\tilde{e}) = \theta_*(e_1) \times M(\eta)\theta_*(e_2) : M(Q_1) \to N_1 \times N_2 ,$$

proving 2.2. □

Now we are in the position to prove the main result of this chapter:

2.3 THEOREM. Let $\eta: Q_1 \to Q_2$ be an isomorphism. Then the following statements are equivalent:

(i) η induces an isoclinism from e_1 to e_2 .

(ii) There exists $\beta': [G_1,G_1] \rightarrowtail\!\!\!\twoheadrightarrow [G_2,G_2]$ with $\beta'\varkappa_1\theta_*(e_1) = \varkappa_2\theta_*(e_2)M(\eta)$.

(iii) $M(\eta)\operatorname{Ker}\theta_*(e_1) = \operatorname{Ker}\theta_*(e_2)$.

PROOF. Condition (ii) makes sense, because $\operatorname{Im}(\varkappa_i\theta_*(e_i))$ is contained in $[G_i,G_i]$, cf. the definition of θ' in (2.1). As $\varkappa_1,\varkappa_2$ are monomorphic, and $\beta',M(\eta)$ are isomorphic, (ii) implies (iii).

Let $\tilde{e}$ be as above and assume (iii). Then 2.2 shows $\operatorname{Ker}\theta_*(\tilde{e}) = \operatorname{Ker}\theta_*(e_1)$. This equation reads in terms of the epimorphisms $(\sigma_i,\tau_i,\gamma_i)\colon \tilde{e} \twoheadrightarrow e$ from (1.2) as follows:

$$M(\gamma_i)\operatorname{Ker}\theta_*(\tilde{e}) = \operatorname{Ker}\theta_*(e_i) \ , \ i=1,2 \ .$$

From 2.1, we obtain that these epimorphisms are isoclinic, and (i) follows from 1.11.

Let $(\eta,\beta')\colon e_1 \sim e_2$. Then 1.11 and 1.8 imply $\beta' = \tau_2'\tau_1'^{-1}$, where $\tau_i' = \tau_i|_{[\tilde{G},\tilde{G}]}$. Diagrams (1.2) and the naturality of (2.1) yield $\beta'\varkappa_1\theta_*(e_1) = \varkappa_2\theta_*(e_2)M(\eta)$. □

2.4 COROLLARY. (i) Let Acl(e) be the group of autoclinisms defined in 1.5. Then

$$\operatorname{Acl}(e) \cong \{ \eta \mid \eta \in \operatorname{Aut}(Q) \ , \ M(\eta)\operatorname{Ker}\theta_*(e) = \operatorname{Ker}\theta_*(e) \} \ .$$

(ii) If e is a generalized representation group of Q, i.e. $\operatorname{Ker}\theta_*(e) = 0$, then $\operatorname{Acl}(e) \cong \operatorname{Aut}(Q)$.

(iii) Any two generalized representation groups of a given group are isoclinic. □

REMARK. Corollary 2.4 (iii) is implicitly contained in HALL [1], for other proofs see GRUENBERG [1] and JONES/WIEGOLD [2].

In terms of projective representations, 2.3 reads as follows:

2.5 COROLLARY. Let $\gamma\colon Q_1 \to Q_2$ be an isomorphism. Then γ induces an isoclinism from e_1 to e_2 , if and only if for all (not necessarily finite dimensional) projective representations P of Q_2 over some algebraically closed field of characteristic zero, the following holds: P can be lifted in e_2 , precisely if $P\gamma$ can be lifted in e_1 .

PROOF. The projective representation $P\gamma$ can be lifted in e_1 , if and only if P can be lifted in

$$e' = (\gamma^{-1})^* e_1 \; : \; N_1 \rightarrowtail^{\varkappa_1} G_1 \xrightarrow{\gamma\pi_1} Q_2 \; .$$

Furthermore we have $\theta_*(e_1) = \theta_*(e_1')M(\gamma)$. Hence, the corollary follows from II.2.20 and 2.3. □

In the sequel, let us fix the group Q and consider an isoclinism class Φ, which contains central extensions whose factor groups are isomorphic to Q. Then Φ has a representative of the form $e: N \rightarrowtail G \twoheadrightarrow Q$, and by 2.3 Φ determines at least one subgroup of $M(Q)$, namely $\mathrm{Ker}\ \theta_*(e)$. In fact each extension in Φ is isomorphic to an extension of the form above. On the other hand, let U be any subgroup of $M(Q)$. Then I.3.8 implies the existence of a extension e' with $\theta_*(e')=\nu: M(Q) \twoheadrightarrow M(Q)/U$ (natural projection). In particular we have $\mathrm{Ker}\ \theta_*(e') = U$, and e' is a stem extension. If we combine these observations with 2.3 and the naturality of θ_* , we easily obtain the following results of P. Hall:

2.6 PROPOSITION. (i) Each central extension is isoclinic to a stem extension. In particular, each group is isoclinic to a stem group.

(ii) The isoclinism classes of central extensions with factor groups isomorphic to Q correspond to the orbits of $\mathrm{Aut}(Q)$ on the set of subgroups of $M(Q)$ with respect to the action $U \longmapsto M(\gamma)U$, $U \subseteq M(Q)$, $\gamma \in \mathrm{Aut}(Q)$. □

2.7 PROPOSITION. Let $\gamma: Q_1 \to Q_2$ be an isomorphism. Then the following conditions are equivalent:

(i) There exists a subgroup U of N_1 such that γ induces an isoclinism from e_1/U: $N_1/U \rightarrowtail G_1/\varkappa_1 U \twoheadrightarrow Q_1$ to e_2.

(ii) $M(\gamma)\operatorname{Ker}\theta_*(e_1) \subseteq \operatorname{Ker}\theta_*(e_2)$.

If (ii) holds, the possible subgroups U of N_1 are given by the condition $\operatorname{Im}(\theta_*(e_1)) \cap U = \theta_*(e_1)M(\gamma^{-1})\operatorname{Ker}(\theta_*(e_2))$.

PROOF. By 2.3, condition (i) holds, if and only if $\operatorname{Ker}(\theta_*(e_1/U)) = M(\gamma^{-1})\operatorname{Ker}(\theta_*(e_2))$. Let ν be the projection from N_1 onto N_1/U, then $\theta_*(e_1/U) = \nu\theta_*(e_1)$, as θ_* is natural. Hence, the equality above is equivalent to

$$M(\gamma^{-1})\operatorname{Ker}\theta_*(e_2) = \{ x \mid x \in M(Q_1),\ \theta_*(e_1)x \in U \},$$

which holds precisely when

$$\operatorname{Im}\theta_*(e_1) \cap U = \theta_*(e_1)M(\gamma^{-1})\operatorname{Ker}\theta_*(e_2)$$

and

$$M(\gamma^{-1})\operatorname{Ker}\theta_*(e_2) \supseteq \operatorname{Ker}\theta_*(e_1),$$

and we are done. □

2.8 REMARKS. (i) Let e be a generalized representation group of Q. Then the extensions e/U for $U \subseteq N$ represent (in general not uniquely) all isoclinism classes with factor groups isomorphic to Q.

(ii) Assume that e is a gen. representation group of Q. Then $\varkappa\theta_*(e): M(Q) \to \varkappa N \cap [G,G]$ is isomorphic. Let $\eta \in \operatorname{Aut}(Q)$. By 2.3 resp. 2.4 it induces an autoclinism (η,ξ) of e. The definitions of isoclinism, $M(\eta)$, and $\theta_*(e)$ yield that $\xi'\varkappa\theta_*(e) = \varkappa\theta_*(e)M(\eta)$, where ξ' is the restriction of ξ to $\varkappa N \cap [G,G]$. Hence, we have the action $U \longmapsto \xi'U$ of Acl(e) on the subgroups of $\varkappa N \cap [G,G]$, which is equivalent to the action of Aut(Q) on the subgroups of M(Q) from 2.6, and again the orbits correspond uniquely to the isoclinism classes with factor group Q. If $(U_i')_{i\in I}$ are representatives of these orbits, the isoclinism classes are represented by $(e/U_i)_{i\in I}$, $U_i = \varkappa^{-1}U_i'$.

(iii) Assume that e_1 is the centralization of a free presentation of Q_1. In this case we have $N_1 \supseteq M(Q_1)$, and $\theta_*(e_1)$ is the embedding of $M(Q_1)$ into N_1. Then the situation of 2.7 is slightly more comfortable. Let $\gamma\colon Q_1 \to Q_2$ be any isomorphism. Then we have a morphism $(\alpha,\beta,\gamma)\colon e_1 \to e_2$ with $\operatorname{Ker}\beta = \varkappa_1 \operatorname{Ker}\alpha$, and $\alpha|_{M(Q_1)} = \theta_*(e_2)M(\gamma)$. Let $U = \operatorname{Ker}\alpha$. Then (α,β,γ) induces a monomorphism $(\alpha',\beta',\gamma)\colon e_1/U \rightarrowtail e_2$, which is isoclinic by 2.1. If e_2 is a stem extension, then (α',β',γ) is even isomorphic.

The proof of 2.8 is left as an easy exercise. □

A group Q is called capable, if there exists a group G with $G/Z(G) \cong Q$. These groups will be studied in more detail in Chapter IV. Nevertheless we can show here, how isoclinism can be used for a first characterization.

2.9 PROPOSITION. Let $e\colon N \overset{\varkappa}{\rightarrowtail} G \overset{\pi}{\twoheadrightarrow} Q$ be a generalized representation group of Q. Then the following conditions are equivalent:

(i) Q is capable.

(ii) $\varkappa N = Z(G)$.

PROOF. Obviously (ii) implies (i). Assume (i). Then there exists an extension $e'\colon N' \overset{\varkappa'}{\rightarrowtail} G' \twoheadrightarrow Q$ such that $\varkappa' N' = Z(G')$. By 2.8,(i) there exists a subgroup U of N such that e/U is isoclinic to e'. Proposition 1.4 implies that N/U is mapped onto the center of $G/\varkappa U$, and we obtain $\varkappa N = Z(G)$. □

2.10 REMARK. Let Q be capable. Then families of groups exist, having Q as central factor group. By 2.9 (the middle groups of) the representation groups of Q also represent a unique family with Q as central quotient, and all other families with this property are represented by certain factor groups of a representation group. P. Hall called the family, which contains the representation groups of Q, the maximal family.

We close this section with a characterization of the groups isoclinic to finite groups.

2.11 PROPOSITION (KING [1]). Let G be a group. Then the following properties are equivalent:

(i) G is isoclinic to a finite group.

(ii) $G/Z(G)$ is finite.

(iii) G is isoclinic to a finite subquotient of itself.

PROOF. The implications (iii) $\Rightarrow$ (i) $\Rightarrow$ (ii) are trivial. Assume that $Q = G/Z(G)$ is finite. Then there exists a finitely generated subgroup G_1 of G with $G = Z(G)G_1$. Thus G_1 is isoclinic to G by 1.7,(iv). Look at $e_1\colon Z(G_1) \hookrightarrow G_1 \twoheadrightarrow Q$. As a central subgroup of finite index in a finitely generated group, $Z(G_1)$ is finitely generated abelian. Decompose $Z(G_1) = T \times A$, where T is the (finite) torsion subgroup and A is torsionfree. As $M(Q)$ is finite by II.3.10, $Z(G_1) \cap [G_1,G_1] = \mathrm{Im}\ \theta_*(e_1)$ is also finite, hence lies in the torsion part, and $A \cap [G_1,G_1] = 0$. It follows from 1.17,(ii) that $G_1 \twoheadrightarrow G_1/A$ is an isoclinic epimorphism. Thus G is isoclinic to its subquotient G_1/A, and G_1/A is finite as an extension of $Z(G_1)/A \simeq T$ by Q. □

3. The Isomorphism Classes of Isoclinic Central Extensions and the Hall Formulae

Throughout this section we consider central extensions

$$e_i : N_i \overset{\varkappa_i}{\rightarrowtail} G_i \twoheadrightarrow Q_i \ .$$

From the results in Section 1 we know that the following subquotients are invariants of isoclinism:

Q_i resp. $G_i/\varkappa_i(N_i)$, $[G_i,G_i]$,

$G_i/([G_i,G_i]\varkappa_i(N_i))$, $\varkappa_i(N_i) \cap [G_i,G_i]$,

whereas the subquotients

$$[G_i,G_i]\varkappa_i(N_i)/[G_i,G_i] \cong \varkappa_i(N_i)/(\varkappa_i(N_i) \cap [G_i,G_i]) \cong$$
$$\cong \text{Coker } \theta_*(e_i) =: B_i$$

may differ for isoclinic extensions. The situation is illustrated by the following diagram:

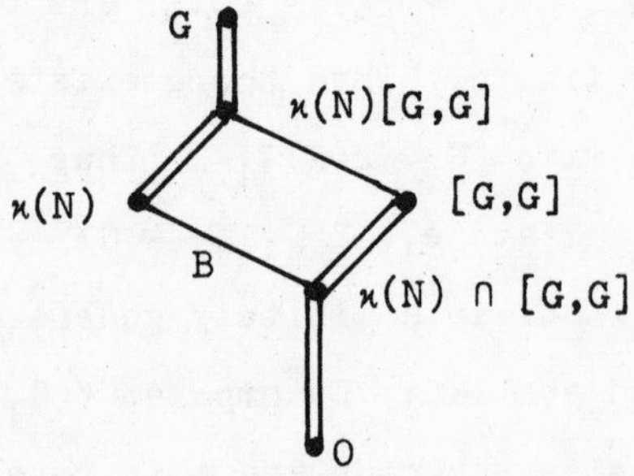

The abelian group B_i is called the "branch factor group" of e_i . It vanishes if and only if e_i is a stem extension and its order (in case of finite groups) is called the "branch factor". Because of 1.4 it makes sense to call $Z(G)/(Z(G) \cap [G,G])$ the branch factor of a group G. It is the aim of this section to solve the following problem: We assume that an isoclinism class is given by a stem extension e_o, and that B is an abelian group. Then representatives of the isomorphism classes (in the sense of I.(1.4)) of those central extensions e_i will be determined, which are isoclinic to e_o and whose branch factor groups B_i are isomorphic to B. If we restrict our attention to isoclinism of groups, i.e. to extensions e_i with $\varkappa_i(N_i) = Z(G_i)$, we obtain (up to isomorphism of groups) the groups in an isoclinism family with a given branch factor group. The isomorphism classes can be worked out by dividing the given isoclinism class in certain subclasses, which are given by either one of the following equivalence relations:

3.1 DEFINITION. (i) Let $(\eta,\xi): e_1 \sim e_2$ be an isoclinism and $\alpha: N_1 \to N_2$ an isomorphism. We call $[\eta,\xi,\alpha]$ a strong isoclinism (of the first kind), if $\varkappa_2\alpha$ and $\beta\varkappa_1$ coincide on $\varkappa_1^{-1}(\varkappa_1 N_1 \cap [G_1,G_1])$; e_1 and e_2 are then called strongly isoclinic (of the first kind).

(ii) Let $(\eta,\xi): e_1 \sim e_2$, and $\alpha': (G_1)_{ab} \rightarrowtail\!\!\!\twoheadrightarrow (G_2)_{ab}$. We call $\{\eta,\xi,\alpha'\}$ a strong isoclinism of the second kind, if α' induces an isomorphism from $G_1/(\varkappa_1 N_1\cdot[G_1,G_1])$ to $G_2/(\varkappa_2 N_2\cdot[G_2,G_2])$, which coincides with the one induced by η; e_1 and e_2 are then called strongly isoclinic of the second kind.

For stem extensions the notions of isoclinism and strong isoclinism of both kinds coincide. In general the strong isoclinisms of the first kind from e_1 to e_2 are in one-to-one correspondence with the pairs $[\eta,\delta]$, where δ is an isomorphism from $\varkappa_1 N_1[G_1,G_1]$ to $\varkappa_2 N_2[G_2,G_2]$ whose restrictions α to $\varkappa_1 N_1$ and ξ to $[G_1,G_1]$ have the property that $[\eta,\xi,\alpha]$ is a strong isoclinism (of the first kind). Similarily the strong isoclinisms of the second kind correspond to the pairs $\{\epsilon,\xi\}$, where ϵ is an isomorphism from

$G_1/(\varkappa_1 N_1 \cap [G_1,G_1])$ to $G_2/(\varkappa_2 N_2 \cap [G_2,G_2]$ inducing isomorphisms η, α' such that $\{\eta, \xi, \alpha'\}$ is a strong isoclinism of the second kind.

It is clear from 3.1 how to define strong isoclinism of groups. These equivalence relations for groups were called by P. Hall "the situation of the centers", resp. "the situation of the commutator quotients". The groups G_1 and G_2 are strongly isoclinic of the second kind, if and only if the corresponding stem extensions

(3.1) $\quad Z(G_i) \cap [G_i,G_i] \hookrightarrow G_i \twoheadrightarrow G_i/(Z(G_i) \cap [G_i,G_i])$, $i=1,2$,

are isoclinic. The only-if-part of this statement holds for arbitrary central extensions.

Both kinds of strong isoclinism can be used in order to classify groups or extensions, and it is more or less a matter of taste which notion one prefers. In the following, strong isoclinism always means strong isoclinism of the first kind. Later on we briefly outline how to deal with the second kind in the case of isoclinic groups.

It is obvious that a strong isoclinism $[\eta, \xi, \alpha]$ from e_1 to e_2 is uniquely determined by η and α. The existence of a suitable isomorphism ξ: $[G_1,G_1] \rightarrowtail\!\!\!\twoheadrightarrow [G_2,G_2]$ will be described in the following

3.2 PROPOSITION. Let α: $N_1 \rightarrowtail\!\!\!\twoheadrightarrow N_2$ and η: $Q_1 \rightarrowtail\!\!\!\twoheadrightarrow Q_2$. Then the following properties are equivalent:

(i) η and α induce a strong isoclinism from e_1 to e_2 ,

(ii) $\theta_*(e_2)M(\eta) = \alpha\theta_*(e_1)$.

PROOF. By 2.3, η induces an isoclinism from e_1 to e_2 , if and only if there exists ξ: $[G_1,G_1] \rightarrowtail\!\!\!\twoheadrightarrow [G_2,G_2]$, such that $\xi\varkappa_1\theta_*(e_1) = \varkappa_2\theta_*(e_2)M(\eta)$. If $[\eta, \xi, \alpha]$ is a strong isoclinism, we have in addition $\xi\varkappa_1 = \varkappa_2\alpha$ on $\varkappa_1^{-1}(\varkappa_1 N_1 \cap [G_1,G_1])$. Both conditions above are equivalent to (ii), and we are done. □

3.3 COROLLARY. (i) Let $[e_1],[e_2] \in \mathrm{Cext}(Q,N)$. Then e_1 and e_2 are strongly isoclinic by a strong isoclinism of the form $[1_Q,\xi,1_N]$, if and only if $[e_1] - [e_2]$ lies in $\Psi(\mathrm{Ext}(Q_{ab},N))$, cf. I.3.8.

(ii) Let $[e_1] \in \mathrm{Cext}(Q,N)$. Then e_1 and e_2 are strongly isoclinic, if and only if e_2 is isomorphic to an extension $e_3 \in \mathrm{Cext}(Q,N)$ with

$$[e_3] \in \Psi(\mathrm{Ext}(Q_{ab},N)) + [e_1] .$$

PROOF. Property (i) follows from 3.2 and I.3.8, and (ii) is a consequence of (i) and the naturality of θ_* . □

Let $e \in \mathrm{Cext}(Q,N)$ and denote by $A(e)$ the set of all strong isoclinisms from e to e (called <u>strong autoclinisms</u>). Obviously, $A(e)$ constitutes a group, and each strong isoclinism from e to an extension e_1 induces an isomorphism from $A(e)$ to $A(e_1)$. Hence, we also denote this group by $A(\Phi)$ where Φ is the corresponding class of strongly isoclinic extensions. By 3.3 (ii) the extensions in Φ are represented (up to isomorphism) by a single coset

$$\Omega := [e] + \Psi\,\mathrm{Ext}(Q_{ab},N) .$$

From 3.2 and 3.3 (i) it follows that a pair $(\eta,\alpha) \in \mathrm{Aut}(Q) \times \mathrm{Aut}(N)$ induces a strong autoclinism of e, if and only if it induces one of each e_1 with $[e_1] \in \Omega$. So, having fixed the coset Ω (which is not uniquely determined by Φ), we can identify $A(\Phi)$ with the group of all pairs (η,α) as above. Thus we have an action of $A(\Phi)$ on the representing set Ω by

$$(3.2) \qquad (\eta,\alpha)[e_1] := [\alpha e_1 \eta^{-1}] ,$$

which is the restriction of the well-known action of $\mathrm{Aut}(Q) \times \mathrm{Aut}(N)$ on $\mathrm{Cext}(Q,N)$.

3.4 PROPOSITION. (i) Let $(\eta,\alpha) \in \mathrm{Aut}(Q) \times \mathrm{Aut}(N)$. The following properties are equivalent:

(a) $(\eta,\alpha) \in A(\Phi)$,

(b) $(\eta,\alpha)[e_1] \in \Omega$ for some $[e_1] \in \Omega$,

(c) $(\eta,\alpha)\Omega = \Omega$.

(ii) Let $[e_i] \in \Omega$, where i runs through some set I. Then the following properties are equivalent:

(a) $(e_i)_{i\in I}$ is a system of representatives of the isomorphism classes of extensions in Φ.

(b) The system $([e_i])_{i\in I}$ represents the orbits of $A(\Phi)$ on Ω.

PROOF. (i) The group $\Psi(\mathrm{Ext}(Q_{ab},N))$ is fixed by the action of $\mathrm{Aut}(Q) \times \mathrm{Aut}(N)$ on $\mathrm{Cext}(Q,N)$, which proves the equivalence of (b) and (c). If $(\eta,\alpha) \in A(\Phi)$, we have by 3.2(ii)

$$\theta_*(e) = \alpha\theta_*(e)M(\eta^{-1}) = \theta_*(\alpha e\eta^{-1}) .$$

Now I.3.8 implies $(\eta,\alpha)[e] \in \Omega$. The converse follows analogously.

(ii) By I.1.10, the isomorphism classes of central extensions of N by Q are given by the orbits of $\mathrm{Aut}(Q) \times \mathrm{Aut}(N)$ on $\mathrm{Cext}(Q,N)$. In order to determine the isomorphism classes of extensions in Φ, we can restrict ourselves to the elements in Ω. Hence, the equivalence of (a) and (b) follows from (i). □

In the following we consider a fixed stem extension

$$e_o : N_o \overset{\varkappa}{\rightarrowtail} G \overset{\pi}{\twoheadrightarrow} Q ,$$

and an abelian group B. Furthermore we assume that

$$d_i : N_o \overset{\lambda_i}{\rightarrowtail} N_i \overset{\sigma_i}{\twoheadrightarrow} B \ , \quad i \in I$$

runs through a system of representatives of $Ext(B,N_o)$, and we obtain the following extensions

$$(3.3) \qquad e(d_i) = \lambda_i e_o : N_i \overset{\varkappa_i}{\rightarrowtail} G_i \overset{\pi_i}{\twoheadrightarrow} Q \ ,$$

which will play an important role in the determination of the isomorphism classes in the isoclinism class of e_o . As e_o is a stem extension, each autoclinism (η,ξ) of e_o yields an automorphism $\alpha' = \varkappa^{-1}\xi\varkappa$ of N_o , cf. 1.6. By Γ we denote the isoclinism class of e_o , and by $Acl(\Gamma)$ the subgroup of $Aut(Q) \times Aut(N_o)$, which consists of pairs (η,α') , where η and α' are as above. Obviously , $Acl(\Gamma)$ is isomorphic to $Acl(e)$ for all extensions e that are isoclinic to e_o . For each element $(\delta,\alpha') \in Aut(B) \times Aut(N_o)$, we have the obvious action on $Ext(B,N_o)$, which maps $[d_i]$ onto $(\delta,\alpha')[d_i] := [\alpha' d_i \delta^{-1}]$. This gives rise to an operation of $Acl(\Gamma) \times Aut(B)$ on $Ext(B,N_o)$ by

$$(3.4) \qquad ((\eta,\alpha'),\delta)[d_i] := (\delta,\alpha')[d_i] \ .$$

3.5 PROPOSITION. (i) All extensions $e(d_i)$ from (3.3) are isoclinic to e_o.

(ii) The branch factor group of each $e(d_i)$ is isomorphic to B.

(iii) Each central extension which is isoclinic to e_o and has a branch factor group isomorphic to B, is strongly isoclinic to some $e(d_i)$.

(iv) The extensions $e(d_i)$ and $e(d_j)$ are strongly isoclinic, if and only if $[d_i]$ and $[d_j]$ are conjugate under the action of $Acl(\Gamma) \times Aut(B)$.

PROOF. We have $e(d_i) = \lambda_i e_o$, where $\lambda_i : N_o \to N_i$ is injective. Thus it follows $\theta_*(e(d_i)) = \lambda_i \theta_*(e_o)$, and

$\operatorname{Ker} \theta_*(e(d_i)) = \operatorname{Ker} \theta_*(e_o)$, proving (i) by 2.3.

As e_o is a stem extension, $\theta_*(e_o)$ is an epimorphism, and it follows $\operatorname{Coker} \theta_*(e(d_i)) = \operatorname{Coker} \lambda_i \cong B$, proving (ii).

Let $e_1 \colon N_1 \overset{\varkappa_1}{\rightarrowtail} G_1 \twoheadrightarrow Q_1$ be a central extension with a branch factor group isomorphic to B, and $(\eta, \xi)\colon e_1 \sim e_o$. Hence we have (cf. 2.3):

$$\xi \varkappa_1 \theta_*(e_1) = \varkappa \theta_*(e_o) M(\eta) \quad \text{and} \quad \operatorname{Coker} \theta_*(e_1) \cong B .$$

Let $\lambda_1 := \varkappa_1^{-1} \xi^{-1} \varkappa$, which is a monomorphism from N_o to N_1 , and we obtain

$$\theta_*(e_1) = \lambda_1 \theta_*(e_o) M(\eta) .$$

As $\theta_*(e_o)$ is epimorphic, we have

$$\operatorname{Im} \lambda_1 = \operatorname{Im} \theta_*(e_1) \quad \text{and} \quad \operatorname{Coker} \lambda_1 = \operatorname{Coker} \theta_*(e_1) \cong B .$$

Thus we obtain an extension $d\colon N_o \overset{\lambda_1}{\rightarrowtail} N_1 \twoheadrightarrow B$, and we have $[d] = [d_i]$, for a suitable index $i \in I$. Thus we have an isomorphism $\alpha\colon N_1 \rightarrowtail\!\!\!\!\twoheadrightarrow N_i$ satisfying

$$\lambda_i = \alpha \lambda_1 : N_o \rightarrowtail N_i .$$

Summarizing the results above, we obtain

$$\theta_*(e(d_i)) M(\eta) = \lambda_i \theta_*(e_o) M(\eta) = \alpha \lambda_1 \theta_*(e_o) M(\eta) = \alpha \theta_*(e_1) .$$

By 3.2(ii) , η and α induce a strong isoclinism from e_1 to $e(d_i)$, proving (iii).

It is easy to see that the automorphisms of Q, which induce the isoclinisms from $e(d_i)$ to $e(d_j)$ coincide with those appearing in $\operatorname{Acl}(\Gamma)$. Hence, assertion (iv) is an easy consequence of 3.1 and I.1.10. □

3.6 REMARK. Propositions 3.4 and 3.5 determine the extensions resp. groups in an isoclinism class resp. family up to isomorphism ; 3.5 yields representatives of the classes of strongly isoclinic extensions, whereas 3.4 yields representatives of the isomorphism classes.

The classification of isoclinic groups can also be worked out by using strong isoclinism of the second kind: Let S be a stem group , and consider

$$s : [S,S] \hookrightarrow S \twoheadrightarrow S_{ab} .$$

As above, let B be an abelian group, and let

$$d_i' : B \hookrightarrow N_i \xrightarrow{\tau_i} S_{ab}$$

be representatives of $\mathrm{Ext}(S_{ab},B)$. By T_i we denote the middle group of the backward induced extension $\tau_i^* s$. By 1.9 these groups are exactly those, which map epi-isoclinic onto S with branch factor group B. Then one can show that the groups T_i yield representatives of the classes under strong isoclinism of the second kind, among those groups, which are isoclinic to S and have B as branch factor group. Each class will be obtained once, by again using a suitable action of $\mathrm{Acl}(\Gamma) \times \mathrm{Aut}(B)$. As mentioned above, groups are strongly isoclinic of the second kind, if and only if the corresponding stem extensions (3.1) are isoclinic. For stem extensions, isoclinism and strong isoclinism means the same. Hence, one can again use 3.4 to determine the isomorphism classes. So both kinds of isoclinism might be useful to classify groups in terms of isoclinism. For central extensions with a fixed factor group, strong isoclinism of the second kind does not seem to be as comfortable as the first kind, because the use of backward induced extensions fixes the kernels of extensions, but not the factor groups.

Now we return to the situation of Proposition 3.4. For each $[e] \in \Omega$, we denote by $St[e]$ the stabilizer of $[e]$ in $A(\Phi)$ and by $C(e)$ the group of all automorphisms of e having the form

$(1_N, \delta, 1_Q)$; this is a normal subgroup of Aut(e) .

3.7 LEMMA. We have

(i) $St[e] \cong Aut(e)/C(e)$

(ii) $C(e) \cong Hom(Q_{ab}, N)$

(iii) $C(e) \cong Ext(Q_{ab}, N)$, if Q and N are finite.

PROOF. By 3.4, St[e] is also the stabilizer of [e] in $Aut(Q) \times Aut(N)$, acting on Cext(Q,N) . Hence, assertion (i) follows from I.1.10. Let $e: N \overset{\kappa}{\rightarrowtail} G \overset{\pi}{\twoheadrightarrow} Q$. Then one easily sees that the automorphisms $(1_N, \delta, 1_Q)$ of e are given by $\delta = \{g \longmapsto g \cdot (\kappa f \pi(g))\}$, where f is a homomorphism from Q to N, and different homomorphisms yield different automorphisms, proving (ii). Assertion (iii) is trivial by (ii), see also 4.5. □

An interesting situation is given, if Φ contains an extension e whose automorphisms induce all elements of $A(\Phi)$, i.e. $St[e] = A(\Phi)$. In this case the isomorphism classes of extensions in Φ are in one-to-one correspondence with the orbits of $A(\Phi)$ on $\Psi(Ext(Q_{ab}, N))$. A more detailed consideration of this situation will be given in Section 4. For finite extensions we now can prove the following "Hall-Formula":

3.8 PROPOSITION. Let Q and N be finite, and $e_1, e_2, \ldots, e_n$ representatives of the isomorphism classes in Φ. Then we have

$$\frac{1}{|A(\Phi)|} = \sum_{i=1}^{n} \frac{1}{|Aut(e_i)|} .$$

PROOF. By 3.4 and orbit decomposition, we have

$$|Ext(Q_{ab}, N)| = |\Omega| = \sum_{i=1}^{n} |A(\Phi) : St[e_i]| .$$

Now 3.7 implies

$$|A(\Phi):St[e_i]| = |A(\Phi)|\cdot|C(e_i)|\cdot|Aut(e_i)|^{-1}$$

and

$$|C(e_i)| = |Ext(Q_{ab},N)| ,$$

and the proposition easily follows. □

If we restrict ourselves in 3.8 to the situation where Φ coincides with the class of stem groups of a family Γ, we obtain Hall's well-known formula

$$(3.5) \qquad \frac{1}{|Acl(\Gamma)|} = \sum_{i=1}^{n} \frac{1}{|Aut(G_i)|} ,$$

where $G_1,G_2,\dots,G_n$ form a complete system of non-isomorphic stem groups in Γ. Similar formulae can be obtained, if one restricts 3.8 to Hall's situation of the centers resp. commutator quotients, i.e. strong isoclinism of groups of the first resp. second kind. Another Hall-Formula turns out to be a consequence of 3.5.

3.9 PROPOSITION. (i) The stabilizer $St([d_i])$ of $[d_i]$ under the action (3.4) of $Acl(\Gamma) \times Aut(B)$ is isomorphic to $A(\Phi_i)/U_i$, where Φ_i is the class of central extensions which are strongly isoclinic to $e(d_i)$, and U_i is a normal subgroup of $A(\Phi_i)$, which is isomorphic to $Hom(B,N_o)$.

(ii) If Q, B, and N_o are finite, and $\Phi_1,\dots,\Phi_m$ are the (different) classes under strong isoclinism, and $e_1,\dots,e_k$ representatives of the isomorphism classes of extensions in Γ with branch factor group B, then

$$\frac{1}{|Acl(\Gamma) \times Aut(B)|} = \sum_{i=1}^{m} \frac{1}{|A(\Phi_i)|} = \sum_{j=1}^{k} \frac{1}{|Aut(e_j)|} .$$

PROOF. Let $(\eta,\alpha) \in A(\Phi_i) \subseteq Aut(Q) \times Aut(N_i)$. Then α induces $\alpha' \in Aut(N_o)$, $\delta \in Aut(B)$, and we have $((\eta,\alpha'),\delta) \in Acl(\Gamma) \times Aut(B)$.

Hence we have a homomorphism

$$\varphi : A(\Phi_i) \longrightarrow \mathrm{Acl}(\Gamma) \times \mathrm{Aut}(B) ,$$

and 3.4 and I.1.10 imply

$$\mathrm{Im}\ \varphi = \mathrm{St}([d_i])$$

Denote by U_i the kernel of φ. It consists of all pairs $(1_Q,\alpha)$ such that α induces 1_{N_o} and 1_B. As in 3.7(ii) we obtain $U_i \cong \mathrm{Hom}(B,N_o)$, proving (i).

The first equation in (ii) follows by the same argument as 3.8. Replacing $|A(\Phi_i)|^{-1}$ by the sum of 3.8 (first Hall-Formula), we obtain the second equation. □

4. On Presentations of Isoclinic Groups

The considerations of this section are based on those of Section 3, some of which will appear again in a different guise. We start with a short review of the group $\mathrm{Ext}(Q_{ab},A)$ and of the homomorphism

$$\Psi : \mathrm{Ext}(Q_{ab},A) \rightarrowtail \mathrm{Cext}(Q,A)$$

from the Universal Coefficient Theorem I.3.8. Let

$$(4.1) \qquad \tilde{e} : X \hookrightarrow Y \twoheadrightarrow Q_{ab}$$

be a free presentation of Q_{ab} (as an abelian group, i.e. Y is free abelian). Then the elements of $\mathrm{Ext}(Q_{ab},A)$ are represented by the induced extensions $\alpha\tilde{e}$, where $\alpha \in \mathrm{Hom}(X,A)$; and $[\alpha\tilde{e}] = [\beta\tilde{e}]$, if and only if $\alpha - \beta$ can be extended to Y. The homomorphism Ψ was defined by

$$\Psi[e'] := ab^*[e'] \quad , \quad ab : Q \twoheadrightarrow Q_{ab} .$$

In this section we consider the following situation. Let

$$(4.2) \qquad e : R \hookrightarrow F \xrightarrow{\pi} Q$$

be a central extension by the group Q, and A,B,C subgroups of F satisfying the following conditions:

$$A \subseteq R \ , \quad B \trianglelefteq F \ , \quad B \cap R = A \ , \quad C = RB \ ,$$

$$C/R = [F/R,F/R] \ ,$$

and

$$C/B \hookrightarrow F/B \xrightarrow{\pi'} Q_{ab}$$

is a free (abelian) presentation of Q_{ab} . In addition we have $B/A = [F/A,F/A]$. This situation is illustrated by the following diagram:

(4.3) 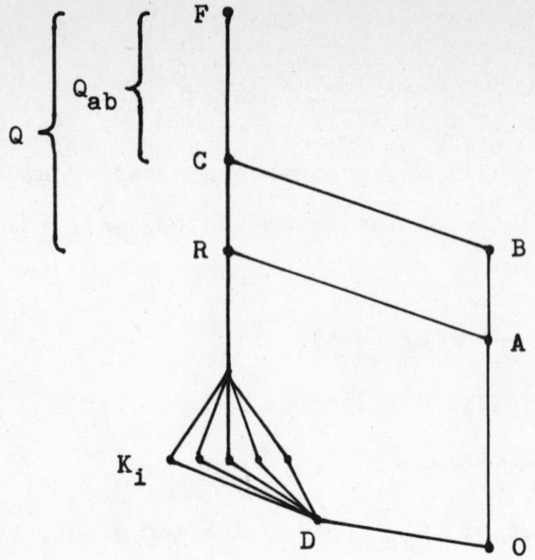

As F/B is free abelian, the same holds for $R/A \cong C/B$, which implies that A has complements in R. For each complement K_i we have corresponding projections

$$a_i \colon R \twoheadrightarrow A \ , \quad k_i \colon R \twoheadrightarrow K_i \quad \text{with} \quad r = a_i(r)\cdot k_i(r) \ .$$

For any two complements K_o, K_1 we obtain

$$f = f(K_o,K_1) = \{Ar \longmapsto a_1(r)\cdot a_o(r)^{-1}\} \ : \ R/A \longrightarrow A \ .$$

Obviously we can regard f also as a homomorphism from C/B to A via the isomorphism $\{Ar \longmapsto Br\}$ from R/A to C/B. If we fix K_o, we thus obtain a one-to-one correspondence between the complements K_i and $\mathrm{Hom}(C/B,A)$. Let $\{ c_j \mid j \in J \}$ be a basis of the free abelian group K_o, then $\{ b_j = Bc_j \mid j \in J \}$ is a basis of C/B, and the complements of A in R are given by

$$\langle c_j\cdot f(b_j) \mid j \in J \rangle \ ,$$

where f runs through the maps from $\{b_j\}$ to A. For each K_i, we denote by λ_i the isomorphism from R/K_i to A defined by

$\lambda_i(K_i r) := a_i(r)$, and we obtain the extension $e(K_i) := \lambda_i(e/K_i)$,

$$\begin{array}{cccccc}
e/K_i & : & R/K_i & \hookrightarrow & F/K_i & \twoheadrightarrow & Q \\
 & & \downarrow{\scriptstyle \lambda_i} & & \downarrow & & \| \\
e(K_i) & : & A & \rightarrowtail & E(K_i) & \twoheadrightarrow & Q
\end{array} .$$

Now we assume that in (4.1) we have $X = C/B$ and $Y = F/B$. Then we can show

4.1 PROPOSITION. Let K_o, K_1 be complements of A in R, and $f = f(K_o, K_1)$. Then

$$[e(K_1)] - [e(K_o)] \;=\; \Psi\,[f\tilde{e}] \;=\; ab^*[f\tilde{e}] ,$$

i.e. the extensions $e(K_i)$ represent a coset of $\Psi(\mathrm{Ext}(Q_{ab},A))$ in $\mathrm{Cext}(Q,A)$; in particular $[e(K_o)] = [e(K_1)]$ holds, if and only if $f(K_o,K_1)$ can be extended from C/B to F/B .

PROOF. Using I.2.4 and putting $e_i := e(K_i)$, one can verify that $[e_1] - [e_o]$ is represented by the following extension:

$$A \overset{\mu}{\rightarrowtail} (A \times F/A)/M \overset{\sigma}{\twoheadrightarrow} Q ,$$

with $M = \{ (f(Ar)^{-1}, Ar) \mid r \in R \}$, $\mu = \{a \longmapsto M(a,1)\}$, $\sigma = \{M(a,Ax) \longmapsto \pi(x)\}$. Let $G = (A \times F/A)/M$, and $\tilde{G} = (A \times F/B)/\tilde{M}$ with $\tilde{M} = \{ (f(Ar)^{-1}, rB) \mid r \in R \}$. Hence we obtain a commutative diagram:

$$\begin{array}{ccccccc}
\tilde{e} & : & C/B & \hookrightarrow & F/B & \overset{\pi'}{\twoheadrightarrow} & Q_{ab} \\
 & & \downarrow{\scriptstyle f} & & \downarrow{\scriptstyle \gamma} & & \| \\
 & & A & \overset{\nu}{\rightarrowtail} & \tilde{G} & \overset{\pi''}{\twoheadrightarrow} & Q_{ab} \\
 & & \| & & \uparrow & & \uparrow{\scriptstyle ab} \\
e_1 - e_o & : & A & \overset{\mu}{\rightarrowtail} & G & \overset{\sigma}{\twoheadrightarrow} & Q
\end{array}$$

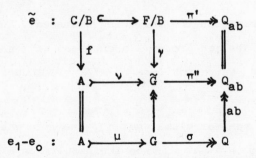

with $\gamma = \{Bx \longmapsto (1,Bx)\tilde{M}\}$, $\nu = \{a \longmapsto \tilde{M}(a,1)\}$, π' and π'' are induced by $\pi \colon F \twoheadrightarrow Q$. This shows that the row in the middle represents $[f\tilde{e}]$, whereas the last row now represents $\Psi[f\tilde{e}]$, and we are done. □

By 3.3 the extensions $e(K_i)$ represent a class of strongly isoclinic extensions of A by Q. Hence the situation above might be helpful, when one tries to classify groups in terms of isoclinism. In addition to 4.1 we can show, c.f. 3.4:

4.2 PROPOSITION. (i) The extensions $e(K_o)$ and $e(K_1)$ are equivalent, if and only if there exists an automorphism of F that centralizes A and F/A and maps K_o to K_1 .

(ii) Let A be finite, F finitely generated, and D the intersection of all complements K_i . Then $e(K_o)$ and $e(K_i)$ are isomorphic, if and only if there exists an automorphism of $e/D: R/D \rightarrowtail F/D \twoheadrightarrow Q$, which maps K_o/D to K_1/D and normalizes $(A \times D)/D$.

PROOF. (i) The extensions $e(K_o)$ and $e(K_1)$ are equivalent, if and only if $f = f(K_o,K_1)$ can be extended from C/B to F/B . If $\tilde{f}$ is an extension of f, we obtain a suitable automorphism α of F by $\alpha(x) := x\tilde{f}(Bx)^{-1}$. On the other hand, each α corresponds by 3.7(ii) to a homomorphism from F/A to A which induces an extension of f.

(ii) The proof of (ii) is based on the following statement: Let Y be a finitely generated free abelian group, X a subgroup of Y and m a positive integer. Then each automorphism α of Y/X is induced by an automorphism of Y/mX .

Proof: Consider the natural homomorphisms $nat_1: Y \twoheadrightarrow Y/X$ and $nat_2: Y/mX \twoheadrightarrow Y/X$, and let $\varphi := \alpha \circ nat_1$. From the results in GASCHÜTZ [2] we obtain the existence of $\beta: Y \twoheadrightarrow Y/mX$ with $nat_2\beta = \varphi$. But then it readily follows that $\mathrm{Ker}\ \beta = mX$. Hence β

yields an automorphism of Y/mX , which induces α.

Let $c_1,\ldots,c_n$ be a basis of K_o , and m the exponent of A. Then $mc_1,\ldots,mc_n$ is a basis of D. If we put $Y = F/B$, $X = C/B$, we obtain $mX = (D \times B)/B$, and we can apply the statement above: Assume that $e(K_o)$ and $e(K_1)$ are isomorphic. Then we have an isomorphism δ: $F/K_o \rightarrowtail\!\!\!\twoheadrightarrow F/K_1$, which maps R/K_o to R/K_1 . As $F/C = (F/R)_{ab}$, δ induces an automorphism α of F/C, which is induced by an automorphism β of $F/(D \times B)$. Because of the obvious isomorphisms

$$F/D \cong F/K_o \curlywedge F/(D \times B) \cong F/K_1 \curlywedge F/(D \times B) ,$$

where in both cases in the fibre products above the groups F/C are identified, δ and β give rise to a suitable automorphism of F/D which maps K_o/D to K_1/D . □

We shall see in the following examples that 4.2(ii) can be false, if we allow A to be infinite.

4.3 EXAMPLES. (i) The Ext-group. Assume that Q is abelian and $R' \hookrightarrow F' \twoheadrightarrow Q$ is a free (abelian) presentation of Q. Then $F := F' \times A$, $R := R' \times A$, $B := A$, $C := R$ satisfy the conditions above. We put $K_o := R'$, whereas we assume that K_1 runs through all complements of A in R. Propositions 4.1 and 4.2 yield a description of Ext(Q,A) in terms of the extensions $e(K_1)$. If we assume $Q = \mathbb{Z}/5$ and $A = \mathbb{Z}$, the group D from 4.2(ii) vanishes, and we readily see that the statement of 4.2(ii) does not hold in this situation. One can also see that this description of Ext(Q,A) can be given for modules over arbitrary rings.

(ii) Representation groups. Let e: $R' \hookrightarrow F' \twoheadrightarrow Q$ be a free presentation of the (arbitrary) group Q, and c(e): $R \hookrightarrow F \twoheadrightarrow Q$ its centralization. We put $A = M(Q) = (R' \cap [F',F'])/[R',F']$,

$B = [F',F']/[R',F']$, and $C = BR$. Then the extensions $e(K_i)$ are the representation groups of Q, and 4.1, 4.2 yield a refinement of II.3.4, 3.5.

(iii) Finite isoclinic p-groups. Let p be a prime, and G a finite stem group, whose order is a power of p. Hence G has a presentation of the following form:

Generators:

x_i, y_i, z_k

Relators:

(a) $[z_k,z_{k'}]$, (b) $[z_k,y_j]$, (c) $[z_k,x_i]$

(d) $z_k{}^p \cdot u_k(z_1,\ldots,z_{k-1})$ (i.e. u_k is a word in $z_1,\ldots,z_{k-1}$)

(e) $y_j{}^p \cdot v_j^1(y_1,\ldots,y_{j-1}) \cdot v_j^2(z)$, $z = (z_1,z_2,\ldots.)$

(f) $[y_{j_1},y_{j_2}] \cdot v^3_{j_1,j_2}(y_1,\ldots,y_{j_1}) \cdot v^4_{j_1,j_2}(z)$, $j_1 < j_2$

(g) $x_i{}^{p^{n_i}} \cdot w_i^1(y) \cdot w_i^2(z)$

(h) $[x_i,y_j] \cdot w^3_{i,j}(y_1,\ldots,y_j) \cdot w^4(z)$

(i) $[x_{i_1},x_{i_2}] \cdot w^5_{i_1,i_2}(y) \cdot w^6_{i_1,i_2}(z)$.

We assume : $[G,G] = \langle y_i,z_k \rangle$, $Z(G) = \langle z_k \rangle$. Furthermore we consider an abelian group P with a basis $a_1,a_2,a_3,\ldots$, where $|a_l| = p^{m_l}$. Let H be the group generated by the elements x_i,y_j,z_k,a_l with the relators (a) - (i) and the following ones:

(j) $[a_e,x_i]$, (k) $[a_e,y_j]$, (l) $[a_l,z_n]$, (m) $[a_l,a_{l'}]$,

(n) $a_l{}^{p^{m_l}} \cdot r_l(z)$.

From the results in Section 3 we see that for any choice of the words $r_l(z)$ in (n), H is isoclinic to G having P as branch factor group, and one obtains representatives of all classes under strong isoclinism. For a fixed H we consider the free group F' with the generators a_l, x_i, y_j, z_k . Let S be the normal closure of the elements of F' given by (a) to (n), i.e. $H \cong F'/S$. By $\tilde{S}$ we denote the normal closure of the words (a) - (f) and (h) - (n) . Let

$$F = F'/\tilde{S}\ ,\quad C = \langle S,a_l,y_j,z_k\rangle/\tilde{S}\ ,$$

$$R = \langle S,a_l,z_k\rangle/\tilde{S}\ ,\quad B = \langle \tilde{S},a_l,y_j,z_k\rangle/\tilde{S}\ ,$$

$$A = \langle \tilde{S},a_l,z_k\rangle/\tilde{S}\ ,\quad K_o = S/\tilde{S}\ .$$

Then F,C,B,R,A,K_o satisfy the conditions of Propositions 4.1, 4.2. A basis of K_o is given by the relators (g) ; and presentations of all groups that are strongly isoclinic to H are given by (a) - (f), (h) - (n), and

(g') $x_i^{p^{n_i}} \cdot w_i^1(y) \cdot w_i'(a) \cdot w_i''(z)$,

where w_i' and w_i'' are arbitrary words in a, resp. z. Similar results can be worked out for arbitrary finite groups. A very rough summary of the observations above is the following: If a group is given by power and commutator relations,one obtains the other isoclinic groups by keeping the commutator relations invariant and changing certain power relations.

4.4 REMARK. Let us assume the situation of 4.2(ii) , i.e. Q is finitely generated and A is finite. Let L denote the subgroup of Aut(e/D) which normalizes $(D \times A)/D$. By 4.2(ii) the orbits of L on the group K_i/D correspond to the isomorphism classes of extensions in the class Φ of extensions which are strongly isoclinic to the extensions e/K_i , resp. $e(K_i)$. We are going to show that the action of L "coincides" with the action of $A(\Phi)$ on the corresponding coset $\Omega = \{[e(K_i)]\}$ in Cext(Q,A) , cf. 3.4. Each element of L induces an automorphism η of Q and an automorphism α of $A \cong (D \times A)/D$. Hence, we have a homomorphism

φ: $L \to (\mathrm{Aut}(Q) \times \mathrm{Aut}(A)$, while 3.4 and the proof of 4,2(ii) yield

(4.4) $\quad \mathrm{Im}\ \varphi = A(\Phi)$.

On the other hand, we have the surjection

$\sigma = \{K_i/D \longmapsto [e(K_i)]\} : \quad \{K_i/D\} \twoheadrightarrow \Omega \subseteq \mathrm{Cext}(Q,A)$,

and we readily see the following relation for all $x \in L$:

(4.5) $\quad \varphi(x)\sigma(K_i/D) = \sigma(x(K_i/D))$.

Now we restrict our attention to finite abelian groups (written additively). Let

$$A = \bigoplus_{i=1}^{u} \langle a_i \rangle , \quad T = \bigoplus_{j=1}^{v} \langle t_j \rangle ,$$

where the numbers $n_i = |a_i|$ and $m_j = |t_j|$ are powers of primes. Let F' be free abelian with a basis $b_1, b_2, \ldots, b_v$, and let R' be the subgroup generated by $r_1 = m_1 b_1, \ldots, r_v = m_v b_v$, so we have a free presentation

(4.6) $\quad R' \overset{\varepsilon}{\hookrightarrow} F' \overset{\pi}{\twoheadrightarrow} T , \quad \pi = \{b_j \longmapsto t_j\}$.

Let $y_{ji} = n_i/\gcd(n_i, m_j)$, and define

$$\psi_{ji} : T \to A ; \quad t_j \longmapsto y_{ji} a_i , \; t_k \longmapsto 0 \quad \text{for all } k \neq j .$$

Hence we have:

$$\mathrm{Hom}(T,A) = \bigoplus_{j,i} \langle \psi_{ji} \rangle , \quad |\psi_{ji}| = \gcd(n_i, m_j) .$$

Let

$$\varphi_{ji} : R' \to A ; \quad r_j \longmapsto a_i , \; r_k \longmapsto 0 \quad \text{for } k \neq j .$$

Then

$$\mathrm{Hom}(R',A) = \bigoplus_{j,i} \langle \varphi_{ji} \rangle ,$$

and

$$\tau = \{\varphi_{ji} \longmapsto \psi_{ji}\} : \mathrm{Hom}(R',A) \to \mathrm{Hom}(T,A)$$

is an epimorphism. Now we readily see that

$\mathrm{Ker}\ \tau = \{ \varphi \mid \varphi \in \mathrm{Hom}(R',A)\ ,\ \varphi(r_j) \in m_j A \text{ for all } j \} = \mathrm{Im}\ \varepsilon^*$

where $\varepsilon^*\colon \mathrm{Hom}(F',A) \to \mathrm{Hom}(R',A)$ is induced by ε, and we obtain the exact sequences

$$(4.7)\qquad \mathrm{Hom}(T,A) \xrightarrow{\pi^*} \mathrm{Hom}(F',A) \xrightarrow{\varepsilon^*} \mathrm{Hom}(R',A) \begin{array}{l} \twoheadrightarrow \mathrm{Ext}(T,A) \\ \twoheadrightarrow \mathrm{Hom}(T,A) \end{array} .$$

On $\mathrm{Hom}(T,A)$ we have the following action of $\mathrm{Aut}(T) \times \mathrm{Aut}(A)$:

$$(\alpha,\beta)\gamma = \beta\gamma\alpha^{-1}\ ,\quad \alpha \in \mathrm{Aut}(T)\ ,\quad \beta \in \mathrm{Aut}(A)\ ,\quad \gamma \in \mathrm{Hom}(T,A)\ .$$

From the considerations above, we can deduce the following

4.5 LEMMA. Let T and A be finite abelian groups. Then $\mathrm{Ext}(T,A)$ and $\mathrm{Hom}(T,A)$ are $(\mathrm{Aut}(T) \times \mathrm{Aut}(A))$-isomorphic. □

In the following we consider the case $T = Q_{ab}$. By 4.1 - 4.5 we have a close connection between the extensions $e(K_i)$, given by the complements K_i , the corresponding coset in $\mathrm{Cext}(Q,A)$, and the groups $\mathrm{Hom}(Q_{ab},A)$. In the very easy situation of Example 4.3(i), we can use 4.5 to evaluate the isomorphism classes of abelian extensions: Let Q and A be finite abelian groups, and $\gamma_1,\ldots,\gamma_n$ representatives of the $(\mathrm{Aut}(Q) \times \mathrm{Aut}(A))$-orbits on $\mathrm{Hom}(Q,A)$. Via the map τ from (4.7) we obtain corresponding elements $f_i \in \mathrm{Hom}(C/B,A)$, cf. 4.3,(i). Define the groups K_i by $f(K_o,K_i) = f_i$. Then 4.1 - 4.5, and (4.4), (4.5) yield that $e(K_1),\ldots,e(K_n)$ represent the isomorphism classes in $\mathrm{Ext}(Q,A)$.

This observation on finite abelian groups can be generalized to certain classes Φ of strongly isoclinic extensions. Let

$$e\ :\ A \rightarrowtail G \twoheadrightarrow Q$$

be a central extension where A and Q are finite and Φ the class of strongly isoclinic extensions containing e. Furthermore we assume that the stabilizer $\mathrm{St}[e]$ of $[e]$ in $A(\Phi]$ satisfies

(4.8) $\quad St[e] = A(\Phi)$,

i.e. the automorphisms of e induce all strong autoclinisms, cf. 3.7. Under this assumption, the action of $A(\Phi)$ on $\Psi\,Ext(Q_{ab},A)$ is equivalent to the action on $\Omega = [e] + \Psi\,Ext(Q_{ab},A)$.

4.6 PROPOSITION. Assume that the finite central extension e satisfies (4.8). Then the isomorphism classes of extensions in Φ are in one-to-one correspondence with the conjugacy classes of elements of Aut(e) that are contained in C(e) . (For the definition of C(e), see 3.7.)

PROOF. The isomorphism classes in Φ correspond to the orbits of $A(\Phi)$ on Ω by 3.5, and by our assumption, Ω can be replaced by $\Psi\,Ext(Q_{ab},A)$. The group $A(\Phi)$, which is a subgroup of $Aut(Q) \times Aut(A)$ induces a subgroup U of $Aut(Q_{ab}) \times Aut(A)$, and it readily follows that the orbits of U on $Ext(Q_{ab},A)$ correspond to the orbits of $A(\Phi)$ on $\Psi\,Ext(Q_{ab},A)$. By 4.5, we can replace $Ext(Q_{ab},A)$ by $Hom(Q_{ab},A)$, which is naturally isomorphic to C(e), and it can be shown that the action of U on $Hom(Q_{ab},A)$ corresponds to the conjugation with the elements of Aut(e) . □

Let us again consider (4.3), and Φ and e as above. Without loss of generality we can assume that the complement K_o of A in R is chosen such that $e = e(K_o)$. As in the abelian case we consider representatives $\gamma_1,\dots,\gamma_n \in Hom(Q_{ab},A)$ of the orbits under the action of U, cf. proof of 4.6. Let us assume that F, C and B are chosen such that $C/B \hookrightarrow F/B \twoheadrightarrow Q_{ab}$ has the form of (4.6). Then we choose (as in the case of abelian groups) homomorphisms $f_1,\dots,f_n \in Hom(C/B,A)$ corresponding to $\gamma_1,\dots,\gamma_n$. The extensions $e(K_i)$ with $f_i = f(K_o,K_i)$ represent the isomorphism classes in Φ.

4.7 EXAMPLES. (i) Representation groups of finite abelian groups. From Proposition II.4.13 one sees that the representation groups of finite nilpotent groups are the direct products of the representation groups of their Sylow subgroups. Hence, we can restrict our attention to abelian p-groups. Let

$$Q = \mathop{\times}_{i=1}^{n} \mathbb{Z}/p^{k_i} \ , \quad \text{where} \quad k_1 \leq k_2 \leq \ldots \leq k_n \ .$$

From II.(4.7) one can deduce

$$M(Q) \cong \mathop{\times}_{i=1}^{n} \ {}^{n-i}\!\!\times \mathbb{Z}/p^{k_i} \ .$$

Let F be a free group on the free generators $x_1, x_2, \ldots, x_n$, and

$$R = \langle \ x_i^{p^{k_i}} , \ [x_i, x_j] \mid i<j \ \rangle^F \supseteq [F,F]$$

$$\tilde{R} = \langle \ x_i^{p^{k_i}} , \ [[F,F],F], [x_i, x_j]^{p^{k_i}} \mid i<j \ \rangle^F \subseteq R \ .$$

(The upper index F denotes "normal closure".) Hence we have a free presentation

$$R \hookrightarrow F \twoheadrightarrow Q \ ,$$

and let

$$\bar{x}_i = Rx_i, \ \tilde{x}_i = \tilde{R}x_i, \ G = F/\tilde{R}, \ M = \langle \ [\tilde{x}_i, \tilde{x}_j] \mid i \leq j \ \rangle = [G,G] \ .$$

Now it is not very diffucult to show that $e\colon M \hookrightarrow G \twoheadrightarrow Q$ is a representation group of Q; see TAPPE [3,13.1] for details.

LEMMA. If p is an odd prime, each automorphism of Q can be lifted to an automorphism of e.

PROOF. We firstly show that the statement of the lemma holds for endomorphisms of Q. An endomorphism γ of Q is given by a map

$$\bar{x}_i \longmapsto \prod_{j=1}^{n} \bar{x}_j^{p^{m_{ij}}} \ , \quad \text{where} \quad k_i + m_{ij} \geq k_j$$

and by

$$x_i \longmapsto \prod_{j=1}^{n} x_j^{p^{m_{ij}}}$$

we obtain an endomorphism $\tilde{\beta}$ of F, which induces γ on $Q \simeq F/R$. We have

$$\tilde{\beta}[[F,F],F] \subseteq \tilde{R} ,$$

and $[[F,F],F] \subseteq \tilde{R}$ implies

$$\tilde{\beta}([x_i,x_j]^{p^{k_i}}) \in \tilde{R} \quad \text{for all} \quad i < j .$$

As p is odd, G is a regular p-group of class 2, what implies

$$\tilde{\beta}(x_i^{p^{k_i}}) \in \tilde{R} \quad \text{for all } i.$$

Thus we have

$$\tilde{\beta}(\tilde{R}) \subseteq \tilde{R} ,$$

and $\tilde{\beta}$ induces an endomorphism β of $G = F/\tilde{R}$. As $M = [G,G]$, we obtain

$$\beta(M) \subseteq M .$$

This finishes the proof for endomorphisms. Now assume that γ is an automorphism of Q and β any endomorphism of e lifting γ. As

$$M = [G,G] = [G,G] \cap Z(G) ,$$

$\beta|_M$ is uniquely determined and $(\gamma, \beta|_M)$ is an autoclinism of e, cf. 2.4. In particular, we obtain that $\beta|_M$ is an automorphism of M. Hence β must be an automorphism of G. □

As a consequence of this lemma, we can apply 4.5 in order to describe the isomorphism classes of representation groups of Q. These extensions form a class Φ of strongly isoclinic extensions, where $A(\Phi)$ is isomorphic to $Aut(Q)$. So the study of the isomorphism classes "reduces" to the (non-trivial) problem of determining the orbits of $A(\Phi)$ on $Hom(Q,M(Q))$ resp. $Hom(F/R,M)$.

Each $\gamma \in Aut(Q)$ induces a unique $\alpha \in Aut(M(Q))$, such that $(\gamma,\alpha) \in A(\Phi)$. The action on $f \in Hom(Q,M(Q))$ is given by $(\gamma,\alpha)f = \alpha f \gamma^{-1}$. Let $f_1,f_2,\ldots,f_r$ be representatives of these orbits, and

$$f_t(\bar{x}_i) = w_t([\tilde{x}_i,\tilde{x}_j]) \ , \ i < j \ ,$$

a word in the elements $[\tilde{x}_i,\tilde{x}_j]$. We then obtain representatives of the isomorphism classes by replacing the relators $x_i^{p^{k_i}}$ in $\tilde{R}$ by $x_i^{p^{k_i}} w_t([x_i,x_j])$. Using similar methods, the representation groups of $Z/p \times Z/p \times Z/p$ were studied by PLATH [1], who showed that there are $2p+10$ isomorphism types.

(ii) The stem extensions of $Q = Z/p^n \times Z/p^n$. Each stem extension of Q is strongly isoclinic to exactly one of the following groups F/R_m, $m = 1,2,\ldots,n$, where $F = \langle x,y \rangle$ is free, and R_n is the normal closure of $\{ x^{p^n}, y^{p^n}, [F,F,F], [x,y]^{p^m} \}$. We have $Aut(Q) = GL(2,Z/p^n)$, and the elements of $Aut(Q)$ induce on $M(Q) \cong Z/p^n$ the multiplication with the determinant. As in Example (i), we see that the automorphisms of Q can be lifted to automorphisms of F/R_m. Hence, $Aut(Q)$ is isomorphic to the group of strong autoclinisms for all m. The action on $Hom(Q,Z/p^n)$ corresponds to the following action of $G = GL(2,Z/p^m)$ on $Z/p^m \times Z/p^m$:

$$a \longmapsto \det(T)\cdot T^{*}a \ ,$$

for all $T \in G$, and the star denotes the transpose of the inverse matrix. Any two vectors are conjugate under G, if and only if they have the same additive order. Hence we have $m+1$ orbits, i.e. there are (up to isomorphism) $m+1$ stem extensions of Q with the order p^{2n+m}.

(iii) Generalized representation group of dihedral groups with elementary abelian center. Let Q resp. H be the dihedral group of order 2^n resp. 2^{n+1} . By I.3.8(iv) H is a representation group of Q. Now it readily follows that all generalized representation groups of Q of order 2^{n+m} with elementary abelian center are strongly isoclinic to

$$G = H \curlyvee (\overset{m}{\times} \mathbb{Z}/2) \ ,$$

where the center of H and a direct factor of $\overset{m}{\times} \mathbb{Z}/2$ are amalgamated. As the automorphisms of Q can be lifted to automorphisms of H, the determination of groups, which are strongly isoclinic to G, is reduced to the study of the action of the group of strong autoclinisms on $\mathrm{Hom}(G_{ab}, Z(G))$, which is isomorphic to the group M of 2×m-matrices over GF(2) . The action of the autoclinisms corresponds to the action of the matrices

$$\left\{ \begin{pmatrix} 1 & 0 \\ 0 & 1 \end{pmatrix} , \begin{pmatrix} 0 & 1 \\ 1 & 0 \end{pmatrix} \right\}$$

on M from the left, and of

$$\left\{ \left(\begin{array}{c|c} 1 & 0\ldots0 \\ \hline * & \\ \vdots & * \\ * & \end{array}\right) \in GL(m,2) \right\}$$

on M from the right. This yields exactly seven isomorphism types, provided $m \geq 3$.

5. Representations of Isoclinic Groups

In this section we study the connection between representations and isoclinism. In most cases we restrict ourselves to finite groups.

The existence of faithful ordinary irreducible representations resp. faithful p-blocks for finite groups was studied in GASCHÜTZ [1], PAHLINGS [2]. Easy examples of abelian groups show that the existence of faithful irreducibles or blocks is not an invariant of isoclinism in general, but the results mentioned above can be used in order to prove:

5.1 PROPOSITION. Let G and H be finite isoclinic groups, which are not isoclinic to one of their proper factor groups. Then G has a faithful ordinary irreducible representation, resp. a faithful p-block, if and only if the same holds for H.

PROOF. By the results of Gaschütz and Pahlings, G has a faithful irreducible character, resp. p-block, if and only if $S(G)$, the socle of G, resp. $O_{p'}(S(G))$ is generated by a single conjugacy class of G. $S(G)$ is the product of all minimal normal subgroups N of G, and for all such N we have

$$N \subseteq [G,G] \quad \text{or} \quad N \cap [G,G] = 0 \; .$$

As G is not isoclinic to a proper factor group of itself, we always have $N \subseteq [G,G]$, hence

$$S(G) \subseteq [G,G] \quad \text{and} \quad O_{p'}(S(G)) \subseteq [G,G] \; .$$

Now the proposition follows from 1.4(iii), the operator isomorphism of $[G,G]$ and $[H,H]$. □

In Section 2 we obtained the following result, cf. 2.5.

5.2 THEOREM. Let G and H be groups, $\gamma: G/Z(G) \rightarrowtail H/Z(H)$ an isomorphism, and K an algebraically closed field of characteristic 0. Then the following properties are equivalent:

(i) γ induces an isoclinism from G to H.

(ii) Each projective K-representation P of $H/Z(H)$ can be lifted in H, if and only if $P\gamma$ can be lifted in G. □

By the results of Chapter II, it suffices to consider in 5.2(ii) the irreducible projective representations, which are finite dimensional if G and H are finite. Let us now shortly discuss the situation of 5.2 in the case where G and H are finite and K is algebraically closed of characteristic p. If D is an irreducible K-representation of G, each normal p-subgroup, in particular $O_p(G)$ is contained in the kernel of D. This gives rise to the following definition:

5.3 DEFINITION. Let G and H be finite groups, K an algebraically closed field of characteristic p, and $e_p(G)$ the following central extension:

$$e_p(G) : \frac{O_p(G)Z(G)}{O_p(G)} \hookrightarrow G/O_p(G) \twoheadrightarrow \frac{G}{O_p(G)Z(G)}$$

We call G and H <u>p-isoclinic</u>, if and only if there exists an isomorphism

$$\gamma : G/O_p(G)Z(G) \rightarrowtail\!\!\!\!\rightarrow H/O_p(H)Z(H)$$

such that each irreducible projective K-representation P of $H/O_p(H)Z(H)$ can be lifted in $e_p(H)$ precisely when $P\gamma$ can be lifted in $e_p(G)$.

Now it is not very surprising that the following result holds:

5.4 PROPOSITION. The following properties are equivalent:

(i) G and H are p-isoclinic.

(ii) $e_p(G)$ and $e_p(H)$ are isoclinic.

PROOF: Let $Q = G/O_p(G)Z(G)$. As the kernel of $e_p(G)$ is a finite p'-group, we have

$$\mathrm{Ker}\ \theta_*(e_p(G)) \supseteq M(Q)_p .$$

Hence the result follows in the same vein as 5.2 from II.2.20. □

As in 5.4 we assume till the end of this section that G and H are finite groups. Furthermore we assume that G and H are isoclinic by an isomorphism $\gamma: G/Z(G) \rightarrowtail\!\!\!\twoheadrightarrow H/Z(H)$, and that G is a stem group. In the following we try to describe the connection between the irreducible representations of G and H over algebraically closed fields more detailed than in 5.2. For sake of simplicity, we write $Z = Z(G)$, $Y = Z(H)$, and assume $Z \subseteq Y$, and that γ induces the embedding $Z \hookrightarrow Y$. In particular we have $Z = [H,H] \cap Y \subseteq [G,G]$. Now we consider

$$L := G \curlywedge H = \{ (g,h) \mid g \in G,\ h \in H,\ \gamma(gZ) = hY \} .$$

By 1.11 L is isoclinic to G and H, and the projections onto G and H are isoclinic epimorphisms. Our assumptions above yield:

$$Z(L) = Z \times Y\ , \quad Z(L) \cap [L,L] = \{ (z,z) \mid z \in Z \} .$$

Let K be an algebraically closed field of characteristic $p \geq 0$. For each $\lambda \in \mathrm{Hom}(Z,K^*)$ we choose an arbitrary, but fixed extension of λ to Y, denoted by $\mu = \mu(\lambda)$. Define $\lambda^{-1}\mu \in \mathrm{Hom}(Z \times Y,K^*)$ by

$$\lambda^{-1}\mu(z,y) := \lambda(z)^{-1}\mu(y) .$$

Hence we have $Z(L) \cap [L,L] \subseteq \mathrm{Ker}(\lambda^{-1}\mu)$, what proves that $\lambda^{-1}\mu$ can be extended to L. Let $\mu^*=\mu^*(\lambda) \in \mathrm{Hom}(L,K^*)$ be an arbitrary, but fixed extension of $\lambda^{-1}\mu$.

Now we consider an irreducible matrix representation $D: G \to GL(n,K)$ of G. By Schur's Lemma we have

$$D|_Z = \lambda \cdot I_n ,$$

where $\lambda = \lambda(D) \in Hom(Z,K^*)$, and I_n is the identity matrix of rank n. From D we obtain an irreducible representation of L:

$$\tilde{D} = \{(g,h) \longmapsto \mu^*(g,h)D(g)\} : L \to GL(n,K),$$

where $\mu^* = \mu^*(\lambda(D))$ is defined as above. The definition of $\tilde{D}$ yields

$$X := \{ (z,1) \mid z \in Z \} \subseteq Ker(\tilde{D}) .$$

As $L/X \cong H$, we obtain an irreducible representation $\hat{D}$ of H:

(5.1) $\hat{D}(h) := \mu^*(g,h) \cdot D(g)$,

where $g \in G$ is any element with $\gamma(gZ) = hY$, the value of $\hat{D}(h)$ does not depend on the choice of g.

Let $U = Hom(H/Y[H,H],K^*)$ and regard it as a subgroup of $Hom(H,K^*)$. By $\tau_1,\tau_2,\dots,\tau_{b'}$ we denote a transversal to U in $Hom(H,K^*)$, which is obtained by extending each element of $Hom(Y[H,H]/[H,H],K^*)$ to H_{ab} . Thus we have

(5.2) $b' = |Y[H,H]/[H,H]|_{p'} = |Y/Z|_{p'}$,

which is the p'-part of the branch factor of H; b' equals the branch factor if p = 0.

5.5 LEMMA. If λ runs through $Hom(Z,K^*)$, then $\nu_i(\lambda) = \{y \longmapsto \mu(\lambda)(y) \cdot \tau_i(y)\}$ runs through $Hom(Y,K^*)$.

The proof of 5.5 is easy. □

Now we are in the position to write down the desired correspondence between the irreducible K-representations of G and H:

(5.3) $D \longrightarrow \{ \hat{D}\tau_i = \{h \longmapsto \hat{D}(h)\tau_i(h)\} \mid i=1,2,\dots,b' \}$, and it readily follows

(5.4) $\hat{D}\tau_i|_Y = \nu_i(\lambda(D))\cdot I_n$

where ν_i is defined as in 5.5.

5.6 THEOREM. Let D_1 and D_2 be irreducible K-representations of G. Then the following properties are equivalent:

(i) $\hat{D}_1\tau_i$ and $\hat{D}_2\tau_j$ are equivalent representations of H.

(ii) $i = j$, and D_1 and D_2 are equivalent.

PROOF. If D_1 and D_2 are equivalent, then $\lambda(D_1) = \lambda(D_2)$, and the definitions above show that $\hat{D}_1\tau_i$ and $\hat{D}_2\tau_i$ are equivalent, proving that (ii) implies (i). If $\hat{D}_1\tau_i$ and $\hat{D}_2\tau_j$ are equivalent, they induce the same characters of Y, i.e. $\nu_i(\lambda(D_1)) = \nu_j(\lambda(D_2))$. By 5.5 we obtain $i = j$ and $\lambda(D_1) = \lambda(D_2)$, and again the proof follows from the definition of $\hat{D}_1\tau_i$ and $\hat{D}_2\tau_j$. □

5.7 COROLLARY. Let $D_1,D_2,\ldots,D_m$ be a complete system of irreducible K-representations of G. Then $\hat{D}_i\tau_j$, $1 \leq i \leq m$, $1 \leq j \leq b'$ is a complete system of irreducible K-representations of H.

PROOF. By 5.6 the representations $\hat{D}_i\tau_j$ are pairwise inequivalent. The number m equals the number of p'-classes of G and by 1.17 mb' is the number of p'-classes of H, proving the corollary. □

REMARK. In all considerations above, it suffices to assume that K is a common splitting field of G and H.

5.8 COROLLARY. Let H_1 and H_2 be finite isoclinic groups, and let $m_1(n)$ resp. $m_2(n)$ be the number of irreducible K-representations of H_1 resp. H_2 of degree n. Then

$$m_1(n)\cdot|H_2|_{p'} = m_2(n)\cdot|H_1|_{p'} .$$

PROOF. Use 5.7 to compare H_1 and H_2 with a common isoclinic stem group. □

5.9 REMARK. By the results of II.1 we see that the correspondence between the linear irreducible K-representations of G and H also yields a correspondence between the irreducible K-representations of G/Z and H/Y which can be lifted in G resp. H. If we fix transversals to Z in G and to Y in H, different $\lambda \in \mathrm{Hom}(Z,K^*)$ yield inequivalent factor systems of G/Z, whereas for a fixed λ all $\nu_i(\lambda) \in \mathrm{Hom}(Y,K^*)$ yield equivalent factor systems.

We finish this section with a few remarks on the modular representation theory of the isoclinic groups G and H from above. For the moment, we assume that K is an algebraic number field which is a splitting field for G and H; let p be a prime, R a valuation ring in K with maximal ideal P containing p. Furthermore we can assume that the representations above are written over R, and for each representation D, we denote by $\overline{D}$ its reduction modulo P. We also assume in the following that for each $\lambda \in \mathrm{Hom}(Z,K^*)$ the corresponding characters $\mu(\lambda)$ and $\mu^*(\lambda)$ are chosen such that $\overline{\lambda_1} = \overline{\lambda_2}$ implies $\overline{\mu(\lambda_1)} = \overline{\mu(\lambda_2)}$ and $\overline{\mu^*(\lambda_1)} = \overline{\mu^*(\lambda_2)}$. Now we can prove

5.10 PROPOSITION. Under the assumptions just made, the following properties are equivalent:

(i) $\hat{D}_1\tau_i$ and $\hat{D}_2\tau_j$ are in the same p-block of H.

(ii) $\overline{\tau_i} = \overline{\tau_j}$, and D_1 and D_2 are in the same p-block of G.

PROOF. We firstly remark that τ_i and τ_j coincide, if and only if τ_i and τ_j coincide on the p'-part of the branch factor group $Y[H,H]/[H,H]$. The values of the central characters of blocks containing $\hat{D}_1\tau_i$ resp. $\hat{D}_2\tau_j$ on the classes of H which are contained in Y are given by $\overline{\nu_i(\lambda(D_1))}$ resp. $\overline{\nu_j(\lambda(D_2))}$. If $\hat{D}_1\tau_i$ and $\hat{D}_2\tau_j$ belong to the same block, then $\lambda(D_1)$ and $\lambda(D_2)$ must coincide on the p'-part of $Y/([H,H] \cap Y)$, which implies $\overline{\tau_i} = \overline{\tau_j}$. $\hat{D}_1\tau_i$ and $\hat{D}_2\tau_j$ are in the same block, if and only if we have irreducible K-representations $F_1, F_2, \ldots, F_n$, where $F_1 = \hat{D}_1\tau_i$, $F_n = \hat{D}_2\tau_j$, and F_i, F_{i+1} have a common modular irreducible constituent. By 5.7 we can assume $F_k = \hat{D}_k'\tau_{i(k)}$, for suitable irreducible K-representations D_k' of G. The remarks above show that $\tau_i=\tau_{i(1)}$, $\tau_{i(2)}, \ldots, \tau_{i(n)}=\tau_j$ all coincide. Thus $\hat{D}_k'$ and D_{k+1}' have a common modular constituent for all k. The considerations above yield $\overline{\lambda(D_k')} = \overline{\lambda(D_{k+1}')}$ for all k, which implies by our assumption above that $\overline{\mu^*(\lambda(D_k'))} = \overline{\mu^*(\lambda(D_{k+1}'))}$. Hence it follows that D_k' and D_{k+1}' have a common modular constituent, what proves that $D_1 = D_1'$ and $D_2 = D_n'$ are in the same block. Thus (i) implies (ii). The converse follows analogously. □

5.11 COROLLARY. Let K be an algebraically closed field of characteristic $p > 0$. Then the correspondence (5.3) between the irreducible K-representations of G and H can be chosen such that the following properties are equivalent.

(i) $\hat{D}_1\tau_i$ and $\hat{D}_2\tau_j$ are in the same p-block of H.

(ii) $\tau_i = \tau_j$, and D_1 and D_2 are in the same p-block of G.

PROOF. We can assume that the representations given here are the modular constituents of the representations given in 5.10. □

In the following we investigate the connection between 5.10 and 5.11 in more detail. Let K,R,P be as in 5.10, let D be an irreducible K-representation of G written over R, and T a regular matrix over R/P , such that

$$T^{-1}\bar{D}_1 T = \begin{pmatrix} d_1 & & & & * \\ & d_2 & & & \\ & & d_3 & & \\ & & & \ddots & \\ 0 & & & & d_k \end{pmatrix},$$

where $d_1, d_2, \ldots, d_k$ are p-modular irreducible representations of G. (By a theorem in FEIT [1], we can chose K and D such that D is even completely reducible.) Hence we have

$$T^{-1}\hat{D}\tau_i T = \begin{pmatrix} f_1 & & & & * \\ & f_2 & & & \\ & & f_3 & & \\ & & & \ddots & \\ 0 & & & & f_k \end{pmatrix},$$

where $f_j = \hat{d}_j \tau_i$, cf. proof of 5.11. This observation and the results above yield:

5.12 THEOREM. Let G be a finite stem group and H a finite group isoclinic to G; let b be the branch factor of H, b' resp. b_p the p'-part resp. p-part of b, where p is a prime. Each p-block of G corresponds to b' p-blocks of H. If a p-block of G contains n ordinary and m modular irreducible characters, each of the b' corresponding p-blocks of H contains $n \cdot b_p$ ordinary and m modular irreducible characters. If N is the decomposition matrix of G and M the matrix obtained from N by repeating each row b_p-times, then the decomposition matrix of H is given by the block matrix consisting of b' blocks equal to M. □

5.13 EXAMPLE. Let G be the symmetric group of degree 7 and H the non-split extension of $Z/2$ by G , which is isoclinic to G:

$$G = \text{group} \langle x_1,\ldots,x_6 \mid x_i^2 = [x_i,x_{i+j+1}] = (x_i x_{i+1})^3 = 1 \rangle$$

$$H = \text{group} \langle y,x_1,\ldots,x_6 \mid x_i^2 y^{-1} = [x_i,x_{i+j+1}]y^{-2} = (x_i x_{i+1})^3 y^{-1} = 1 \rangle .$$

The decomposition matrices of G for $p = 2$, and 3 are given in KERBER [1], KERBER/PEEL [1], ROBINSON [1]:

p = 2

1				
0	1			
1	1			
1	1	1		
1	0	1		
1	0	1		
1	1	1		
1	1			
0	1			
1				
			1	
			1	1
			2	1
			1	1
			1	

p = 3

1								
1	1							
0	1	1						
1	1	1	1					
0		0	1	1				
1		1	1	1				
1		0		1				
		1						
		1						
					1			
					1	1		
						1		
							1	
							1	1
								1

The branch factor of H equals 2. From 5.12 we obtain the decomposition matrices of H , see next page.

We close this section with the remark that isoclinisms of finite groups yield a correspondence between their defect groups, and also behave nicely with respect to the Brauer correspondence between the groups and their normalizers of the defect groups.

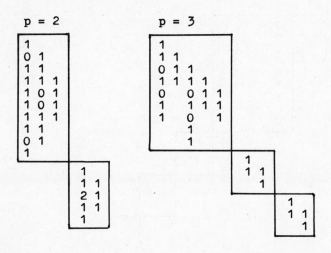

Decomposition matrices of H :

p = 2

1		
1		
0	1	
0	1	
1	1	
1	1	
1	1	1
1	1	1
1	0	1
1	0	1
1	0	1
1	0	1
1	1	1
1	1	1
1	1	
1	1	
0	1	
0	1	
1		
1		

1	
1	
1	1
1	1
2	1
2	1
1	1
1	1
1	
1	

p = 3

1				
1	1			
0	1	1		
1	1	1	1	
0		0	1	1
1		1	1	1
1		0		1
		1		
		1		

1				
1	1			
0	1	1		
1	1	1	1	
0		0	1	1
1		1	1	1
1		0		1
		1		
		1		

1	
1	1
	1

1	
1	1
	1

1	
1	1
	1

1	
1	1
	1

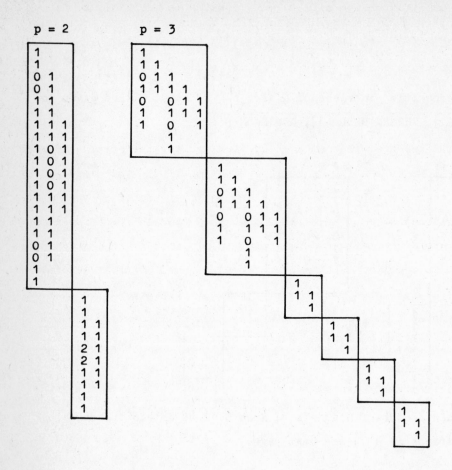

CHAPTER IV. OTHER GROUP-THEORETIC APPLICATIONS OF THE SCHUR MULTIPLICATOR

1. Deficiency of Finitely Presented Groups

Our treatment of the Schur multiplicator used free presentations rather abstractly so far, except for Section II.3. There we saw already that the size of $M(G)$ is severely restricted if the (preferably finite) group G is known to have a "small" free presentation. Here we investigate this relationship closer and also demonstrate that the Schur multiplicator does not contain full information on the possible free presentations of a given group, even if the latter is finite. To this end, we give an elementary treatment of the famous examples of SWAN [1]. Many open problems remain.

The general assumption of this section is that G is a finitely generated group. Equivalently , we assume the existence of a free presentation

(1.1) $$R \hookrightarrow F \xrightarrow{\rho} G$$

with F free of finite rank, say on the basis $x_1,\dots,x_n$. The x_i are called "generators". By the Nielsen-Schreier Theorem, $R = \operatorname{Ker} \rho$ is again a free group (of infinite rank, unless G is finite or $R = 0$). If it is possible to generate R by finitely many elements $r_1,\dots,r_m \in F$ as a normal subgroup of F , then the group G is called finitely presented. In this case the above information is codified as

(1.2) $$G = \langle x_1,\dots,x_n : r_1,\dots,r_m \rangle_\rho$$

and the presentation (in the extended sense) is abbreviated as

(1.3') $P = \langle x_1,\dots,x_n : r_1,\dots,r_m \rangle$, or

(1.3") $P = \langle x_1,\dots,x_n : r_1 = \dots = r_m = 1 \rangle$.

In these formulas F is implicit as the free group on $\{x_1,\dots,x_n\}$ and R as the normal closure of the set of "(defining) relators" r_j and, unless ρ is specified, G is identified with $G(P) := F/R$. The above terminology, most notably the use of the "relations $r_j = 1$" , involves some abuse of language; here we rely on the thorough treatment of this topic in CROWELL/FOX [1; chp. IV].

The following proposition means that the property of "finitely presented" can, in principle, be tested on any finite set of generators. Since this is not used elsewhere in these notes, we record it without proof.

1.1 PROPOSITION (B.H. NEUMANN [1; (8) p. 124]). If the group G has a presentation with n generators and m relators and $Y = \{y_1,\dots,y_k\}$ is any set of generators of G , then G also has a presentation of the form

$$\langle y_1,\dots,y_k : s_1(Y),\dots,s_{k+m}(Y) \rangle . \quad \square$$

1.2 DEFINITION. The <u>deficiency</u> of a finite presentation P as in (1.3') is $\mathrm{def}(P) = m - n$. If G is a finitely presented group, then its <u>deficiency</u> is defined as

$$\mathrm{def}(G) = \inf \{\mathrm{def}(P)\}$$

where P varies over all finite presentations of G . We shall see below that $\mathrm{def}(G)$ is non-negative for finite G and an integer in general. (Warning: In the literature, the deficiency if often defined as the negative of the above.)

1.3 LEMMA. When M is a finitely generated abelian group, let $d(M)$ be the minimum number of generators and $\mathrm{rank}(M) :=$

$= \dim_{\mathbb{Q}}(M \otimes \mathbb{Q}) = d(M/\mathrm{Tor}\, M)$ the rational rank. Then the following holds for finitely generated abelian groups A and B :

(i) $d(A \times B) \leq d(A) + d(B)$;

(ii) $d(A \times B) = d(A) + d(B)$ if B is free-abelian.

PROOF. Easy consequence of the structure theorem for finitely generated abelian groups. □

1.4 PROPOSITION (P. Hall's Inequality). If G is a finitely presented group, then G_{ab} and $M(G)$ are finitely generated abelian and

$$(1.4) \qquad \mathrm{def}\, G \geq d\, M(G) - \mathrm{rank}(G_{ab}) .$$

PROOF (EPSTEIN [1; §1]). Given any finite free presentation P of G as in (1.3'), determing F and R as above. By Proposition I.3.5, or directly by the Schur-Hopf Formula, we have an exact sequence

$$0 \longrightarrow M(G) \longrightarrow R/[R,F] \longrightarrow F_{ab} \longrightarrow G_{ab} \longrightarrow 0 .$$

Now $K := \mathrm{Ker}(F_{ab} \to G_{ab})$, being a subgroup of F_{ab} , is again free-abelian and therefore $R/[R,F] \cong M(Q) \times K$. Since R is generated by elements of the form $fr_if^{-1} = [f,r_i]\cdot r_i$ with $1 \leq i \leq m$ and $f \in F$, the abelian group $R/[R,F]$ is generated by the cosets $r_1[R,F],\ldots,r_m[R,F]$, thus $d(R/[R,F]) \leq m$ and $M(Q)$ is finitely generated. By the additivity of the rank function on short-exact sequences and by Lemma 1.3 (ii), we obtain

$$\begin{aligned} \mathrm{def}(P) = m - n \geq d(R/[R,F]) - n &= dM(G) + d(K) - n \\ &= d(M(G)) + \mathrm{rank}(K) - \mathrm{rank}(F_{ab}) \\ &= d(M(G)) - \mathrm{rank}(G_{ab}) . \quad \square \end{aligned}$$

1.5 REMARKS. (a) As G_{ab} is torsion for a finite group G, in this case P. Hall's Inequality reduces to

$$(1.5) \qquad \operatorname{def}(G) \geq d(M(G)) .$$

But even then the general formula is useful. For we are often given a free presentation without knowing whether the group G defined by it is finite. But one easily computes G_{ab}. If this is finite, P. Hall's Inequality and the comparison with known finite factor groups of G may allow one to prove the finiteness of G. This method was behind the reasoning of Example II.5.8.

(b) Let G be finitely presented. Evidently there exists some presentation P with $\operatorname{def}(P) = \operatorname{def}(G)$. Can one add the requirement that $n = n(P)$ be the minimal number of generators of G? This is an open problem, even for finite groups G. Consult RAPAPORT [1] for results in this direction.

1.6 DEFINITION (EPSTEIN [1]). A finitely presented group is called <u>efficient</u>, if equality holds in (1.4).

B.H. NEUMANN [2] stated (1.5) and raised the problem whether every finite group with trivial multiplicator has zero deficiency. EPSTEIN [1] suspected that $(\mathbb{Z} \times \mathbb{Z}/2) * (\mathbb{Z} \times \mathbb{Z}/3)$ is not efficient. We now give examples of efficient groups and of non-efficient finite groups.

1.7 EXAMPLES. a) According to EPSTEIN [1], every finitely generated abelian group

$$G = \mathbb{Z}/c_1 \times \ldots \times \mathbb{Z}/c_t \times (\mathbb{Z})^r , \quad c_1 | \ldots | c_t$$

is efficient by virtue of its "canonical presentation". This canonical presentation has $n = t + r$ generators, t power relators and $n(n-1)/2$ commutator relators. On the other hand, $\operatorname{rank}(G) = r$ and $M(G)$ has $n(n-1)/2$ cyclic direct summands. (Here we also need that the orders can be arranged in a divisor chain: (n-1) times c_1, (n-2) times $c_2, \ldots,$ r times c_t.) We find

$$\operatorname{def}(G) = \frac{1}{2} n(n-1) - r \geq -1 .$$

The groups $\mathbb{Z}$ and $\mathbb{Z} \times \mathbb{Z}$ are the only abelian groups with deficiency -1 ; a complete list of those with deficiency zero is $\mathbb{Z} \times \mathbb{Z} \times \mathbb{Z}$, 0, $\mathbb{Z}/c \times \mathbb{Z}$ and $\mathbb{Z}/c$ for $c \geq 2$.

b) It will be shown in the next section that all finite metacyclic groups are efficient.

c) The group G of a tame knot with n crossings has a presentation, called the Wirtinger presentation, with n generators $x_1 = x_{n+1}, x_2, \ldots, x_n$ and n relators. The relations express that any two consecutive generators are conjugate in G ; this implies $G_{ab} \cong \mathbb{Z}$. In addition, any relator is known to lie in the consequence of the other (n-1) relators. Thus, by deleting one relator, we obtain a presentation of G with deficiency -1 . Then P. Hall's Inequality gives $d(M(G)) = 0$. Consequently, $M(G) = 0$ and the knot group G is efficient. The necessary facts from knot theory can be found e.g. in CROWELL/FOX [1; chp. VI), where certain "smaller" variants of the Wirtinger presentation are also discussed.

d) The group $SL(2,\mathbb{Z})$ is known to be an amalgated free product of finite cyclic groups such that the presentation

(1.6') $\langle x,y : x^4 = 1 , \ x^2 = y^3 \rangle$

results, where x corresponds to $\begin{pmatrix} 0 & -1 \\ 1 & 0 \end{pmatrix}$ and y to $\begin{pmatrix} 0 & -1 \\ 1 & 1 \end{pmatrix}$ and the superfluous relation $y^6 = 1$ has been deleted. Then the modular group $PSL(2,\mathbb{Z})$ is seen to have the presentation

(1.6") $\langle x,y : x^2 = y^3 = 1 \rangle$.

It follows from (1.6') and (1.6") that the commutator quotient groups are $\mathbb{Z}/12$ and $\mathbb{Z}/6$, respectively. Hence $M(SL(2,\mathbb{Z})) = M(PSL(2,\mathbb{Z})) = 0$ by P. Hall's Inequality and the groups are efficient.

e) Let us look at the infinite dihedral group D_∞ in the same fashion. From the presentation given in Example II.4.7 (b), we find $(D_\infty)_{ab} \cong \mathbb{Z}/2 \times \mathbb{Z}/2$ and we regain the result $M(D_\infty) = 0$ without any effort.

1.8 EXAMPLE. The binary polyhedral groups are, by definition, the non-cyclic finite subgroups of $SU(2)$. Thus they have a faithful 2-dimensional complex representation. Under the celebrated covering

$$\{\pm I\} \hookrightarrow SU(2) \xrightarrow{\tau} SO(3)$$

of topological groups, they correspond to the ordinary polyhedral groups, cf. DUVAL [1; chp. 3]. For $n \geq 2$ the binary dihedral groups of order $4n$ are known to have the presentation

$$\langle s,t : s^2 = t^n = (st)^2 \rangle ,$$

and are commonly called dicyclic. The binary tetrahedral, octahedral, and icosahedral groups have the presentations

$$\langle s,t : s^3 = t^n = (st)^2 \rangle$$

with $n = 3,4,5$, respectively. A simultaneous elementary treatment of these presentations is found in COXETER [1] who departed from problems posed by Threlfall. The above presentation of the binary icosahedral group was already discussed in Example II.5.8. (Actually, we there started with the presentation and showed that its group is a subgroup of $SU(2)$ that maps onto A_5 .) All binary polyhedral groups have trivial multiplicator by virtue of the presentations listed.

We disgress to remark that the concept of deficiency is closely linked with the study of 3-dimensional manifolds. Let M be an oriented compact connected 3-manifold without boundary, with finite fundamental group G , then

$$M(G) \cong H_2(G,\mathbb{Z}) \cong H_2M = 0$$

by results from algebraic topology (e.g. Poincaré duality). EPSTEIN [1; Lemma 2.7] uses this to prove that G has zero deficiency. This explains the previous example since THRELFALL/SEIFERT [1] showed that the binary polyhedral groups are fundamental groups as specified above. (Actually, the real parametrisation

$$\begin{pmatrix} x_0+x_3i \ , & x_2-x_1i \\ -x_2-x_1i \ , & x_0-x_3i \end{pmatrix}, \quad x_0^2+x_1^2+x_2^2+x_3^2 = 1$$

of SU(2) implies that the underlying topological space is a 3-sphere, simply-connected in particular. A finite subgroup G gives rise to a covering

$$SU(2) \longrightarrow M := G\backslash SU(2) = \cup\, Gx \ ,$$

whence M is a 3-manifold as required with fundamental group isomorphic to G.) EPSTEIN [1; Theorem 2.5] proceeds to show that the fundamental groups G of all compact connected 3-manifolds (possibly non-orientable, infinite G allowed) are efficient.

1.9 PROPOSITION (SWAN [1; §2]) Let G be a finitely presented group and U a subgroup of finite index. Then U is finitely presented and

$$(1.7) \qquad \mathrm{def}(U) + 1 \le |G:U|\cdot(\mathrm{def}(G) + 1) \ .$$

PROOF. Let P as in (1.3') be a presentation of G with def G = def P = m − n , let i := |G:U| . Then the Reidemeister-Schreier process yields a presentation of U with $1+(n-1)\cdot i$ generators and $m\cdot i$ relators, see REIDEMEISTER [1] or ROTMAN [1]. Hence

$$\mathrm{def}(U) \le i\cdot m - 1 - (n-1)\cdot i = i\cdot(\mathrm{def}(G)+1) - 1 \ .$$

Alternatively, construct the 2-dimensional CW-complex $C(P)$ with one vertex, n arcs, and m 2-cells. Look at the presentation of U which results from the cell decomposition of the i-fold covering space $C(P)_U$ belonging to the subgroup U of $G \cong \pi_1 C(P)$. □

1.10 EXAMPLES of SWAN [1]. Let $G_k = (\mathbb{Z}/7)^k \rtimes \mathbb{Z}/3$, where a chosen generator t of $\mathbb{Z}/3$ acts on the elementary abelian group $A = (\mathbb{Z}/7)^k$ by squaring, i.e. $tat^{-1} = a^2$ for $a \in A$. (This is an action due to $7 = 2^3 - 1$.) It is asserted that $M(G_k) = 0$ for all k , but $\mathrm{def}(G) > 0$ for $k \geq 3$, actually $\mathrm{def}(G_k) \to \infty$ for $k \to \infty$. These were the first examples of finite groups known to be not efficient.

PROOF. We have $\mathrm{def}\, A = k(k-1)/2$ by Example 1.7 a) and then

$$\mathrm{def}(G_k) \geq \tfrac{1}{3}(\mathrm{def}\, A + 1) - 1 = \tfrac{1}{6}k(k-1) - \tfrac{2}{3} \geq \tfrac{1}{3}$$

for $k \geq 3$ by (1.7). We are left to show the vanishing of $M(G_k)$. Swan's proof of this fact used advanced tools from algebraic topology, the following is elementary. Obviously A is the Sylow 7-subgroup and $\{t\}$ is a Sylow 3-subgroup of G_k . By Proposition I.6.8 $M(G_k)$ has at most 3-torsion and 7-torsion; by Proposition I.6.9 and $M(\mathbb{Z}/3) = 0$, there is no 3-torsion. Invoking I.6.9 now for $p = 7$, we conclude that $\mathrm{res}\colon M(A) \to M(G_k)$ is surjective. Since the isomorphism $M(A) \cong A \wedge A$ of Theorem I.4.7 is natural, t acts on $M(A)$ by $\{m \longmapsto m^4\}$ in view of $a^2 \wedge b^2 = 4(a \wedge b)$ in $A \wedge A$. We know from Proposition I.3.14 (b) that $m^3 = {}^t m \cdot m^{-1} \in$ $\in \mathrm{Ker}(\mathrm{res})$ for all $m \in M(A)$. Finally $\mathrm{res}\colon M(A) \to M(G_k)$ vanishes as $M(A)$ has exponent seven (exponent one only for $k = 1$). □

1.11 PROPOSITION. Let $\pi\colon G \to Q$ be an epimorphism of groups such that G is nilpotent, $N := \mathrm{Ker}\, \pi \subseteq [G,G]$ and $M(Q) = 0$. Then π is an isomorphism.

PROOF. We invoke the 5-term exact sequence I.(3.3) determined by $N \triangleleft G \twoheadrightarrow Q$. The assumptions imply $\mathrm{Ker}(\pi_{ab}) = 0$ and then $N/[N,G] = 0$. Due to the nilpotency of G ,

$$N = [N,G] = [\ldots[N,G]\ldots,G] \subseteq [\ldots[G,G]\ldots,G] = 0 .$$

Of course, this is a special case of the Stallings-Stammbach Theorem, see STALLINGS [1], STAMMBACH [1]. □

This innocent looking proposition can have striking applications as WAMSLEY [2] noticed. Start with a "small" presentation of some "unknown" nilpotent group G . Add further relators such that the corresponding factor group Q has the same commutator quotient, i.e. $Q_{ab} \cong G_{ab}$ in the obvious fashion. If it happens that Q is a known group with trivial multiplicator, we have achieved two things:

(i) the previously unknown group G has been identified as Q ;

(ii) typically, a smaller presentation for the known group Q has been found.

(Warning: Due to Theorem 1.13, finite nilpotent groups on 4 or more generators have non-trivial multiplicator.)

1.12 EXAMPLE (WAMSLEY [2; p.135]). Let p be an odd prime and G be defined by

$$(1.7) \qquad \langle\, a,b \,:\, c := a^{-1}b^{-1}ab,\; c^{-1}ac = a^{1+p},\; cbc^{-1} = b^{1-p} \,\rangle .$$

This group was introduced by MACDONALD [1] who proved it to be finite and nilpotent, but the precise order remained open. Let Q be the factor group subject to the additional relations

$$(1.8) \qquad a^{p^2} = 1 \;,\quad b^{p^2} = 1 \;,\quad c^p = 1 \;.$$

Clearly $G_{ab} \cong Q_{ab} \cong \mathbb{Z}/p \times \mathbb{Z}/p$. It is claimed that Q has order p^5 and trivial multiplicator. This being granted, G turns out to be isomorphic to Q and Q to be efficient by virtue of (1.7).

We are left to prove the claim. In the course of doing so, we exhibit Q as a representation group of the non-abelian group H of order p^3 and exponent p and find $M(H) \cong \mathbb{Z}/p \times \mathbb{Z}/p$. (Another representation group of H was specified by P. HALL [1; p.139].) The relations for Q imply

$$c^k a c^{-k} = a^{1-kp} \quad \text{and} \quad c^k b c^{-k} = b^{1-kp} \quad , \quad k \in \mathbb{Z} .$$

Next $b^{-1}a^p b = (b^{-1}ab)^p = (ac)^p =$

$$= ac^p \cdot c^{-(p-1)} ac^{p-1} \cdot \ \ldots \ \cdot c^{-2}ac^2 \cdot c^{-1}ac = a^p$$

due to $\sum_{i=1}^{p-1} i = \frac{p-1}{2} \cdot p$ and $c^p = a^{p^2} = 1$. Likewise b^p is central. The order of Q cannot be less than p^5. For upon adding either the relation $a^p = 1$ or $b^p = 1$, we obtain a known group of order p^4, viz. group No. 13 (resp. No. 12 for $p = 3$) in the list of HUPPERT [1; Satz III.12.6; Aufg. 29, p.349]; were $|Q| = p^4$ then both $a^p = 1 = b^p$ and thus $Q \cong H$ would follow, contradiction. We conclude

$$|Q| = p^5 , \quad Z(Q) = \{a^p, b^p\} , \quad Q/Z(Q) \cong H .$$

We now invoke the 6-term exact sequence consisting of I.(3.3') and I.(4.2), for the central extension $e_H \colon \mathbb{Z}/p \hookrightarrow H \twoheadrightarrow \mathbb{Z}/p \times \mathbb{Z}/p$. Using $|H_{ab}| = p^2$ and $|M(\mathbb{Z}/p \times \mathbb{Z}/p)| = p$ and counting cardinalities, we conclude that $\theta_*(e_H)$ is bijective and thus $\chi_H = \chi(e_H)$ surjective. Therefore $M(H)$ is a factor group of $\mathbb{Z}/p \times \mathbb{Z}/p$. The corresponding sequence for

$$e_Q : \ Z(Q) \hookrightarrow Q \xrightarrow{\ \pi\ } H \ ,$$

with p-ranks written underneath, is

$$\begin{array}{ccccccccccc} Q_{ab} \otimes Z(Q) & \xrightarrow{\chi_Q} & M(Q) & \longrightarrow & M(H) & \xrightarrow{\theta_*(e_Q)} & Z(H) & \longrightarrow & Q_{ab} & \xrightarrow{\pi_{ab}} & H_{ab} \\ 4 & & ? & & \leq 2 & & 2 & & 2 & & 2 \ . \end{array}$$

We first find that π_{ab} is isomorphic and thus $\theta_*(e_Q)$ surjective -

next that $\theta_*(e_Q)$ is isomorphic. This exhibits e_Q as a representation group of H by Proposition II.2.14 and proves $M(H) \cong \mathbb{Z}/p \times \mathbb{Z}/p$. Hence χ_Q is surjective.

We are going to prove $\chi_Q = 0$ and to this end invoke the explicit definition of χ_Q in terms of the free presentation (1.7) - (1.8) of Q . Since M(Q) has exponent at most p and $R/[R,F] \cong M(Q) \times \{\text{free-abelian}\}$, we conclude

$$(R^p[R,F]) \cap (R \cap [F,F]) = [R,F] .$$

We define elements u, v, w of R by

$$u = a^{p^2} , \quad v = a^{-1-p}c^{-1}ac , \quad w = c^p .$$

Let us abbreviate $\alpha = 1+p$. An easy induction gives

$$\text{(i)} \quad c^{-k}ac^k \equiv a^{\alpha^k}v^{1+\alpha+\ldots+\alpha^{k-1}} \quad \text{mod } [R,F]$$

for $k > 0$. We examine the case $k := p$ and note

$$\eta := \sum_{i=1}^{p-1}(1+\alpha+\ldots+\alpha^{i-1}) \equiv \sum_{i=1}^{p-1} i \equiv 0 \quad \text{mod } p ,$$

$$\xi := \sum_{i=o}^{p-1} \alpha^i = \eta(\alpha-1) + p \equiv p \quad \text{mod } p^2 .$$

Thus $a \equiv w^{-1}aw = c^{-p}ac^p \equiv a^{1+p\xi}v^\xi$ mod $[R,F]$ with $\xi \equiv 0$ mod p and $1 + p\xi \equiv 1 + p^2 \text{ mod } p^3$. Inserting $a^{p^2} = u$, we conclude

$$\text{(ii)} \quad u \equiv 0 \quad \text{mod } R^p[R,F] .$$

Redoing an argument from above, we use (i) and (ii) to deduce

$$\begin{aligned} b^{-1}a^pb &= (b^{-1}ab)^p = (ac)^p \\ &= a\cdot c^p\cdot c^{-(p-1)}ac^{p-1}\cdot \ \ldots \ \cdot c^{-1}ac \\ &\equiv w\cdot a^\xi v^\eta \qquad\qquad \text{mod } [R,F] \\ &\equiv w\cdot a^p \qquad\qquad \text{mod } R^p[R,F] , \end{aligned}$$

$$\text{(iii)} \quad a^{-p}b^{-1}a^pb \equiv w = c^p \quad \text{mod } R^p[R,F] .$$

In the same fashion we obtain (iv) from

$$a^{-1}b^{-p}a = (a^{-1}b^{-1}a)^p = (cb^{-1})^p = cb^{-1}c^{-1}\cdot \ \dots \ \cdot c^{p-1}b^{-1}c^{-(p-1)}\cdot c^p\cdot b^{-1},$$

(iv) $a^{-1}b^{-p}ab^p \equiv w = c^p \mod R^p[R,F]$.

By the direct description I.(4.3) of χ_Q , its image is generated by the cosets of $[a,b^p]$ and $[a^p,b] = [b,a^p]^{-1}$ alone. As χ_Q is bihomomorphic, we have

(v) $[a^p,b^p] \equiv [a,b^p]^p \equiv [a,b^{p^2}] \equiv 0 \mod [R,F]$.

Next, as $cac^{-1}a^{p-1}$ and $cbc^{-1}b^{p-1}$ and $[a^p,b^{-1}]$ lie in R and thus are central in $F/[R,F]$, we find

$$\begin{aligned} a^{-1}b^{-1}ab = c &= c\cdot c\cdot c^{-1} \\ &= [(cac^{-1})^{-1},(cbc^{-1})^{-1}] \\ &\equiv [a^{p-1},b^{p-1}] = a^{-1}a^pb^pb^{-1}a^{1-p}b^{1-p} \\ &\equiv a^{-1}b^pa^pb^{-1}a^{1-p}b^{1-p} \quad \text{by (v)} \\ &= a^{-1}b^p[a^p,b^{-1}]b^{-1}ab^{1-p} \\ &\equiv a^{-1}b^{p-1}ab^{1-p}\cdot[a^p,b^{-1}] \quad \mod [R,F] . \end{aligned}$$

Consequently,

(vi) $[b^{-1},a^p] \equiv [a^{-1},b^p] \mod [R,F]$.

Again, as χ_Q is bihomomorphic, the combination of (iii) - (vi) gives $w^2 \equiv 0$ and then $w \equiv 0 \mod R^p[R,F]$. We conclude $w \in [R,F]$ and finally $\chi_Q = 0$.

It is still not known whether all finite p-groups are efficient. At least, the following famous result implies that p-groups on four or more generators have non-trivial multiplicator and thus need more relators than generators.

1.13 THEOREM (Golod-Šafarevič). Let G be a finite p-group and $d(G)$ its minimal number of generators. Then $d(\mathrm{Cext}(G,\mathbb{Z}/p)) > d(G)^2/4$, consequently

(1.9) $\mathrm{def}(G) \geq d(M(G)) > \frac{1}{4}d(G)^2 - d(G)$.

GOLOD/ŠAFAREVIČ [1] proved $d(Cext(G,\mathbb{Z}/p)) > [d(G)-1]^2/4$. We recommend Roquette's proof which is given in each of the books of GRUENBERG [1], HUPPERT [1], and D.L. JOHNSON [1]. Here we are content to relate (1.9) to the more customary formulations of the Golod-Šafarevič Theorem. By the Burnside Basis Theorem and the Universal Coefficient Theorem I.3.8, we have $d(G) = d(G_{ab})$ and

$$d(Cext(G,\mathbb{Z}/p)) = d(M(G)) + d(G) .$$

This and (1.5) reduce (1.9) to the asserted inequality for $d(Cext(G,\mathbb{Z}/p))$. Choose a presentation $R \lhd F \twoheadrightarrow G$ with F the free group on $d(G)$ generators, let r' be the minimum number of generators for the $\mathbb{Z}[G]$-module R_{ab}/R_{ab}^p . The first and trickiest step is $r' > d(G)^2/4$, see HUPPERT [1; III.18.9]. Next r' is shown to agree with $d(R/R^p[R,F])$, see HUPPERT [1; V.25.2 (c)]. Finally $d(Cext(G,\mathbb{Z}/p)) = d(R/R^p[R,F])$ follows from the exact sequence I.(2.3) associated with the free presentation and the coefficients $\mathbb{Z}/p$, trivial action.

It is only fair to say that our knowledge on deficiency is quite deficient. We draw the reader's attention to WAMSLEY [2]; most questions of this survey are still open. JOHNSON/ROBERTSON [1] listed many finite groups known to have zero deficiency. WIEGOLD [4] recalled conjectures like the one that $A_5 \times A_5$ is not efficient, i.e. requires at least three more relators than generators.

Complete results are available for certain modified concepts. For example, a finite p-group G can be presented with $d(G)$ generators and $d(Cext(G,\mathbb{Z}/p))$ relators as a pro-p-group; this was the starting point in GOLOD/ŠAFAREVIČ [1]. Another good case are nilpotent varieties of exponent zero. For example, let $\mathfrak{B}$ be the variety of all nil-3 groups G , i.e. of nilpotency class at most three. Whenever (1.2) is interpreted as a $\mathfrak{B}$-free presentation of

G , it is understood that R contains $[[F,F],F]$, such relators need not be listed separately. The roles of G_{ab} and $M(G)$ are now played by certain "varietal homology groups" - instances of which will be treated in Section 6. Then the analogue of P. Hall's Inequality turns out to be equality by STAMMBACH [2], [3; Cor. IV.6.6].

2. Metacyclic Groups

A metacyclic group is, by definition, a finite group G which possesses a cyclic normal subgroup A such that G/A is also cyclic. (Cyclic groups G are not excluded, though they constitute a trivial case.) Subgroups and quotient groups of metacyclic groups are again metacyclic.

Metacyclic groups have well-known free presentations which depend on four parameters. However, it is a difficult problem to enumerate the group isomorphism types in terms of these parameters. We propose a new choice for the fourth one - this parameter turns out to be the order of the Schur multiplicator and thus a group invariant. We show that every metacyclic group possesses a metacyclic representation group with trivial multiplicator, among others. We then compute the Schur multiplicator and prove the efficiency of arbitrary metacyclic groups, results first obtained by WAMSLEY [1] in a different fashion. Much of the material of this section is taken from BEYL [1].

2.1 PROPOSITION. Let $Q = \mathbb{Z}/n$ be the cyclic group of order n with generator τ , let A be a Q-module. Then

$$\mu : \operatorname{Opext}(Q,A) \cong \frac{A^Q}{(1+\tau+\ldots+\tau^{n-1})\cdot A} ,$$

the isomorphism μ being specified in the proof.

This result is a special case of the classical extension theory, see SCHREIER [1] or ZASSENHAUS [1; III §7]. For the purposes of 2.2 and 2.5 below, it is given a different proof. Recall from I.(6.4) that A^Q is the subgroup of the fixed elements.

PROOF. Let $\mathbb{Z}$ be the infinite cyclic group, free on the generator b. We invoke the obvious free presentation

$$(2.1) \qquad \bar{e}_n : R \hookrightarrow \mathbb{Z} \xrightarrow{\nu} \mathbb{Z}/n ,$$

of Q, with $\nu(b) = \tau$ and R free on $\sigma = b^n$. Then the group $\mathbb{Z}$ acts on A by ${}^b a = \tau\cdot a := {}^\tau a$ and, by Theorem I.2.7,

$$\theta^*(\bar{e}_n,A) = \{\alpha \longmapsto \alpha_*[\bar{e}_n]\} : \mathrm{Hom}_Q(R,A) \to \mathrm{Opext}(Q,A)$$

is a homomorphism with kernel consisting of all restrictions $\alpha = d|_R$ of derivations $d\colon \mathbb{Z} \to A$. And $\theta^*(\bar{e}_n,A)$ is surjective by Lemma I.2.2; thus $\mathrm{Opext}(Q;A)$ is a quotient group of $\mathrm{Hom}_Q(R,A)$ with $R \cong \mathbb{Z}$.

A homorphism $\alpha\colon R \to A$ of groups is uniquely determined by specifying $u=\alpha(\sigma) \in A$. It is Q-homomorphic precisely when $\tau\cdot\alpha(\sigma) = \alpha({}^\tau\sigma) = \alpha(b^n) = \alpha(\sigma)$. Thus the admissible α's are parametrized by A^Q. If $d\colon \mathbb{Z} \to A$ is any derivation, $\alpha = d|_R$ has $u = d(b^n) = (1+\tau+\ldots+\tau^{n-1})\cdot d(b)$. But every $a \in A$ is of the form $d(b)$, take the derivation defined by

$$d(b^k) = \begin{cases} (1+\tau+\ldots+\tau^{k-1})\cdot a & \text{for } k \geq 1 \\ 0 & \text{for } k = 0 \\ -(\tau^{-1}+\ldots+\tau^{-k})\cdot a & \text{for } k \leq -1 \end{cases} .$$

This is the explicit description of μ^{-1}: Pick a representative element $u \in A^Q$, define $\alpha\colon R \to A$ by $\alpha(\sigma) = u$ and let $\mu^{-1}(u + \text{Denominator}) = \alpha_*[\bar{e}_n]$. □

2.2 COROLLARY. Every metacyclic group has a presentation

$$(2.2) \qquad \langle\, a,b : a^m = 1 ,\ b^n = a^t ,\ bab^{-1} = a^r \,\rangle ,$$

the integers m,n,r,t being subject to the conditions

(2.3a) $\quad m,n > 0$,

(2.3b) $\quad r^n \equiv 1 \mod m$,

(2.3c) $\quad m|t(r-1)$.

Conversely, the group defined by (2.2), subject to the conditions (2.3), is metacyclic of order $m \cdot n$.

This characterization is stated by ZASSENHAUS [1; III §7] and there attributed to Hölder.

PROOF. Let A be a cyclic normal subgroup of G such that G/A is also cyclic, $m = |A|$ and $n = |G/A|$. Choose $b \in G$ that maps on a generator of G/A . Conjugation by b determines an automorphism of A which necessarily has the form $a \mapsto a^r$ for some integer r prime to m (A being generated by a). Condition (2.3b) expresses that A is a G/A-module. Of course $b^n \in A$, thus $b^n = a^t$ for some $t \in \mathbb{Z}$. Condition (2.3c) means $a^t \in A^Q$ and follows from

$$a^t = b^n = bb^nb^{-1} = (bab^{-1})^t = a^{rt} .$$

We now recall from Proposition 2.1 and its proof that $A \rightarrowtail G \twoheadrightarrow G/A$ is forward-induced from $\bar{e}_n$ along $\alpha = \{\sigma \mapsto a^t\}$. According to I.1.5, G is isomorphic to $A \rtimes \mathbb{Z} \,/\, \{ (a^{-kt}, b^{kn}) \mid k \in \mathbb{Z} \}$. Now $A \rtimes \mathbb{Z}$ has the presentation

$$\langle a,b : a^m = 1,\ bab^{-1} = a^r \rangle$$

while the normal subgroup in the denominator accounts for the additional relator $a^{-t}b^n$. (One additional relator suffices since the normal subgroup is cyclic.) On the basis of Proposition 2.1, the converse is just a paraphrase of the facts that the conditions turn $A = \mathbb{Z}/m$ into a $\mathbb{Z}/n$-module and the fixed element a^t specifies a forward-induced extension. □

There is a smallest positive integer t satisfying (2.3c), viz. $t_o = m/(m,r-1)$, and all other such t are integral multiples of t_o . Here (m,n) denotes the (positive) greatest common divisor of the integers m and n , not both zero. We choose $\lambda = t/t_o$ as a defining parameter instead of t, this will be vindecated soon.

2.3 DEFINITION. Let $G(m,n,r,\lambda)$ be the metacyclic group of order $m \cdot n$ presented by (2.2) with $t = \frac{m\lambda}{(m,r-1)}$, the parameters being subject to (2.3a) and (2.3b). Let

$$e(m,n,r,\lambda) : \mathbb{Z}/m \overset{\varkappa}{\rightarrowtail} G(m,n,r,\lambda) \overset{\pi}{\twoheadrightarrow} \mathbb{Z}/n$$

be the supporting metacyclic group extension where $\mathbb{Z}/m = \{a\}$ and $\varkappa$ is inclusion, $\pi(b)$ is the distinguished generator τ of $\mathbb{Z}/n$.

2.4 REMARK. For any abelian group B, consider the endomorphism

$$r^* = \{b \longmapsto b^r\} : B \to B ;$$

this is automatically Q-homomorphic in case B is even a Q-module. Clearly $0^* = 0$ and $1^* = 1$. Write $r = \pm 1 .. \pm 1$ with $|r|$ summands. By Theorem I.2.4, we have $\text{Opext}(Q,(-1)^*) = (-1)^*$ and

$$\text{Opext}(Q,r^*) = r^* : \text{Opext}(Q,B) \to \text{Opext}(Q,B) ;$$

we express this property by saying: "Opext is additive".

2.5 PROPOSITION. Let $Q = \mathbb{Z}/n$ with generator τ, let $A = \mathbb{Z}/m$ with generator a. Assume $r^n \equiv 1 \mod m$ and regard A as a $\mathbb{Z}/n$-module by $\tau \cdot a = a^r$. Then $\text{Opext}(Q,A)$ is cyclic of order

(2.4) $$h(m,n,r) = \frac{(m,r-1)}{m} \cdot (m, 1+r+\ldots+r^{n-1}) .$$

The formula $[e(m,n,r,\lambda)] = \lambda[e(m,n,r,1)]$ exhibits $[e(M,n,r,1)]$ as a generator of $\text{Opext}(Q,A)$.

PROOF. We invoke the description of μ from Proposition 2.1, now $A = \mathbb{Z}/n$. Recall that the typical metacyclic extension is (congruent to) $\alpha\bar{e}_n$ for some Q-homomorphism $\alpha: R \to A$ with

$$\alpha(\sigma) = a^t = a^{\lambda t_o} , \quad t_o = \frac{m}{(m,r-1)} .$$

Let $\alpha_o: R \to A$ be the Q-homomorphism with $\alpha_o(\sigma) = a^{t_o}$. Then $e(m,n,r,1) = \alpha_o(\bar{e}_n)$ and $e(m,n,r,\lambda) = \alpha(\bar{e}_n)$ by the very definition, cf. the proof of 2.2. Since α is the composite

$$R \xrightarrow{\alpha_o} A \xrightarrow{\lambda^*} A \quad ,$$

the formula $[e(m,n,r,\lambda)] = \lambda[e(m,n,r,1)]$ results from Proposition I.1.11 (a) and the additivity of Opext . The formula for $h(m,n,r)$ is immediate from Proposition 2.1. In detail, A^Q is the cyclic group of order m/t_o generated by a^{t_o} and the denominator is the subgroup of order $m/(m,1+r+\ldots r^{n-1})$ generated by $a^{1+r+\ldots+r^{n-1}}$. □

2.6 LEMMA. If $(t,L) = 1$ and $K \neq 0$ for integers K,L,t , then there exists $s \in \mathbb{Z}$ with $(s,K) = 1$ and $s \equiv t \mod L$. Consequently, $\{r+K\mathbb{Z} \longmapsto r+L\cdot\mathbb{Z}\}: (\mathbb{Z}/K)^* \to (\mathbb{Z}/L)^*$ is an epimorphism for $L|K$. Here $(\mathbb{Z}/K)^*$ denotes the unit group of the ring $\mathbb{Z}/K$.

PROOF. Let d be the largest factor of K prime to t . Set $s := t + d\cdot L$. For the primes p dividing K , distinguish the cases $p|t$ and $p \nmid t$. If $L|K$, then $\{r+K\mathbb{Z} \longmapsto r+L\cdot\mathbb{Z}\}$ is a (unital) homomorphism of rings, surjective by the first assertion. Alternatively, this lemma follows from the Chinese Remainder Theorem, see HASSE [1 ; §4 No.2]. □

2.7 THEOREM. Let Q be an arbitrary group and A a finite Q-module. Assume that $\mathrm{Opext}(Q,A)$ is cyclic. Whenever $[e_1]$ and $[e_2]$ have the same order in $\mathrm{Opext}(Q,A)$, there exists an automorphism α of A with $e_2 \equiv \alpha e_1$; in particular, the middle groups G_1 and G_2 are isomorphic.

PROOF. Since A is finite, there is $m \neq 0$ with $m^*=0: A \to A$ in the notation of 2.4. The additivity of Opext implies $m^*=0: \mathrm{Opext}(Q,A) \to \mathrm{Opext}(Q,A)$. Consequently, $\mathrm{Opext}(Q,A)$ is finite cyclic of some order L dividing m . Let $[e_o]$ be a generator, let $[e_1] = s[e_o]$ and $[e_2] = t[e_o]$ have the same order in

$\mathrm{Opext}(Q,A)$. Then $(s,L) = (t,L) = d$, say. As s/d and t/d are both prime to L/d , there exists an integer r_1 prime to L/d with $t/d \equiv r_1 \cdot s/d \mod L/d$. Now Lemma 2.6 gives an integer r prime to m with $r \equiv r_1 \mod L/d$, hence with

$$(m,r) = 1 \quad \text{and} \quad t \equiv r \cdot s \mod L .$$

Let $\alpha = r^*: A \to A$, this is an isomorphism. (There are integers μ, ν with $1 = \mu \cdot m + \nu \cdot r$, then $\nu^* r^* = 1$ due to $m^* = 0$.) Since Opext is additive, we conclude

$$\alpha_*[e_1] = rs[e_o] = t[e_o] = [e_2] . \quad \square$$

We recall from Section I.1 that an isomorphism of extensions is a morphism $(\alpha,\beta,\gamma)\colon e_1 \to e_2$ in which α,β,γ are isomorphic.

2.8 COROLLARY. Assume $r^n \equiv 1 \mod m$, let $h = h(m,n,r)$ as in (2.4). The extensions $e(m,n,r,\lambda)$ and $e(m,n,r,\mu)$ are isomorphic precisely when $(\lambda,h) = (\mu,h)$. The metacyclic group $G(m,n,r,\lambda)$ is also supported by $e(m,n,r,\lambda')$ with $\lambda' = (\lambda,h)$ dividing h .

This corollary is part of more complete results which include the classification of metacyclic extensions up to isomorphism, the details are in BEYL [1].

PROOF. Abbreviate $e_1 = e(m,n,r,\lambda)$ and $e_2 = e(m,n,r,\mu)$. Assume that e_1 and e_2 are isomorphic. By Theorem I.1.10 there are automorphisms α of $\mathbb{Z}/m$ and γ of $\mathbb{Z}/n$ with

$$[e_2] = (\gamma^*)^{-1} \alpha_*[e_1] .$$

(Warning: Implicit in the cited result is the use of an intermediate module structure on $\mathbb{Z}/m$ which is described by a power of r .) Since $(\gamma^*)^{-1}\alpha_*$ is an isomorphism of abelian groups, both extension classes have the same order in $\mathrm{Opext}(\mathbb{Z}/n,\mathbb{Z}/m)$. As the order of

$e(m,n,r,\lambda)$ is $h/(\lambda,h)$ by Proposition 2.5, we conclude $(\lambda,h) = (\mu,h)$. For the converse, first note that $(\lambda,h) = (\mu,h)$ means that $[e_1]$ and $[e_2]$ have the same order. By Theorem 2.7 there is a $\mathbb{Z}/n$-automorphism α of $\mathbb{Z}/m$ such that $[e_2] = \alpha_*[e_1]$. This yields an isomorphism $(\alpha,\cdot,1)\colon e_1 \to e_2$ of extensions. The second assertion follows from the first by letting $\mu := \lambda' = (\lambda,h)$. □

The preceding corollary allows us to impose the condition

(2.5) $\quad \lambda \mid h(m,n,r)$

without loss of generality. This we now do - except that we sometimes prefer $\lambda = 0$ to $\lambda = h(m,n,r)$ in the split case.

2.9 THEOREM. Let $r^n \equiv 1 \mod m$. The metacyclic group $G(m,n,r,1)$ can be presented with two generators and only two relators. In particular, it has trivial Schur multiplicator.

PROOF (BEYL [2]). Choose integers k and l with

$$(m,r-1) = k\cdot m + l(r-1) .$$

Then l is prime to $t := m/(m,r-1)$ and Lemma 2.6 gives an s prime to m with $s \equiv l \mod t$, hence with

(a) $\quad (m,r-1) \equiv s(r-1) \mod m$.

Since $[b,a^s] = a^{s(r-1)}$ in $G(m,n,r,1)$, the following relations (b) and (c) clearly hold:

(b) $\quad b^n = a^t \quad$ with $t = \dfrac{m}{(m,r-1)}$,

(c) $\quad [b,a^s] = a^{(m,r-1)}$.

It is claimed that the relators of the defining presentation of $G(m,n,r,1)$ are consequences of (b) and (c); this is obvious for $b^n = a^t$. We conclude from (b) that a^t commutes with b, and from (c) that $[b,a^s]$ commutes with a . Hence

$$a^{st} = b\,a^{st}b^{-1} = (ba^sb^{-1})^t = ([b,a^s]a^s)^t = [b,a^s]^t a^{st}$$

and $[b,a^s]^t = 1$. Invoking (c), we obtain

(d) $$a^m = a^{(m,r-1)t} = [b,a^s]^t = 1 .$$

Since $(s,m) = 1$, there is $v \in Z$ with $sv \equiv 1 \mod m$. Then

$$\begin{aligned} [b,a] &= [b,a^{sv}] = ba^{sv}b^{-1}a^{-sv} && \text{by (d)} \\ &= (ba^sb^{-1})^v a^{-sv} = ([b,a^s]a^s)^v a^{-sv} \\ &= [b,a^s]^v a^{sv-sv} = a^{v(m,r-1)} && \text{by (c)} \\ &= a^{vs(r-1)} = a^{r-1} && \text{by (d),(a)} . \end{aligned}$$

The final assertion follows from Proposition 1.4 (P. Hall's Inequality). □

2.10 THEOREM (BEYL/JONES [1]). Every metacyclic group has a metacyclic group with trivial multiplicator among its representation groups. If $r^n \equiv 1 \mod m$, then the Schur multiplicator of $G(m,n,r,\lambda)$ is cyclic of order $(\lambda,h(m,n,r))$ with $h(m,n,r)$ as in (2.4).

PROOF. Without loss of generality let the parameters m,n,r,λ be subject to (2.3a), (2.3.b), and (2.5). Since $r-1/(m,r-1)$ is prime to $m/(m,r-1)$, Lemma 2.6 gives an integer t with

$$(t,m) = 1 \quad \text{and} \quad t \equiv \frac{r-1}{(m,r-1)} \mod \frac{m}{(m,r-1)} .$$

Put $s := t(m,r-1) + 1$, then

$$s \equiv r \mod m .$$

Consequently, $(m,r-1) = (m,s-1)$ and $h(m,n,r) = h(m,n,s)$ and $G(m,n,r,\lambda) \simeq G(m,n,s,\lambda)$. Due to (2.5) and (2.4), we have

(2.6) $$m\cdot\lambda \mid m\cdot h(m,n,s) \mid (s-1)(1+s+\ldots+s^{n-1}) = s^n - 1 .$$

Therefore $\tilde{G} := G(m\cdot\lambda,n,s,1)$ is defined as a metacyclic group of order $m\cdot n\cdot\lambda$ on the generators $\tilde{a}$ and $\tilde{b}$, cf. Cor. 2.2. Let $\rho\colon \tilde{G} \to G(m,n,s,\lambda)$ be the epimorphism with $\rho(\tilde{a}) = a$ and $\rho(\tilde{b}) = b$. This ρ is well-defined because the relations of $\tilde{G}$ are respected; the latter is obvious except for the relation with $\frac{m\cdot\lambda}{(m\cdot\lambda,s-1)}$ in the exponent of $\tilde{a}$. However,

$$\lambda \mid h(m,n,s) \mid (m,s-1)$$

and our choice of t imply

$$(m\cdot\lambda,s-1) \mid (m\cdot(m,s-1),s-1) = (m,s-1)(m,t) = (m,s-1) .$$

Counting orders, we conclude that the kernel of ρ coincides with $\{\tilde{a}^m\} \cong \mathbb{Z}/\lambda$. It is claimed that

$$(2.7) \qquad \tilde{e} : \quad \{\tilde{a}^m\} \hookrightarrow \tilde{G} = G(m\cdot\lambda,n,s,1) \xrightarrow{\ \rho\ } G(m,n,s,\lambda)$$

is a representation group of $G(m,n,r,\lambda) \cong G(m,n,s,\lambda)$. First,

$$[\tilde{b},\tilde{a}^m] = \tilde{b}\tilde{a}^m\tilde{b}^{-1}\tilde{a}^{-m} = \tilde{a}^{m(s-1)} = 1$$

shows that $\tilde{e}$ is central. Again by $(m\cdot\lambda,s-1) = (m,s-1)$, the abelianized groups have the same order $n\cdot(m,s-1)$ and coincide. Thus $\tilde{e}$ is a stem extension and $\theta_*(\tilde{e})\colon MG(m,n,s,\lambda) \to \mathbb{Z}/\lambda$ is surjective. Now $\tilde{G}$ has trivial multiplicator by Theorem 2.9. The claim and the remaining assertion now follow by Proposition II.2.14. □

2.11 COROLLARY. All metacyclic groups are efficient.

PROOF. If $G(m,n,r,\lambda)$ has trivial multiplicator, then $(\lambda,h(m,n,r)) = 1$ by Theorem 2.10 and $G(m,n,r,\lambda) \cong G(m,n,r,1)$ by Corollary 2.8, the latter has deficiency zero by Theorem 2.9. For the other metacyclic groups, the presentation (2.2) is efficient. □

2.12 EXAMPLE. Consider the dihedral groups $D_{2m} = G(m,2,-1,0)$ of order $2m$. We easily compute

$$h(m,2,-1) = \begin{cases} 1 & \text{for } m \text{ odd,} \\ 2 & \text{for } m \text{ even .} \end{cases}$$

Thus $M(D_{2m}) = 0$ whenever m is odd. From now on, let m be even. Then our method replaces $\lambda = 0$ by $\lambda = 2$, we conclude $M(D_{2m}) \cong Z/2$. Which representation groups result from Theorem 2.10? If we choose $s = r = -1$, then we obtain the generalized quaternion group $G(2m,2,-1,1)$ as a representation group of D_{2m}. If $4|m$, then the choice $s = m-1$ is also possible and because of $h(2m,2,m-1) = 1$ leads to the so-called quasi-dihedral group

$$G(2m,2,m-1,1) \cong G(2m,2,m-1,0) .$$

2.13 LEMMA. Let $r^n \equiv 1 \mod m$. The center of the metacyclic group $G(m,n,r,\lambda)$ is generated by

$$a^{t_o} \quad \text{and} \quad b^{n_o}$$

where $t_o = m/(m,r-1)$ and n_o is the smallest positive integer ν such that $m|r^\nu-1$. In other words, n_o is the order of $r + mZ$ in $(Z/m)^*$. The central factor group is the metacyclic group $G(t_o,n_o,r,0)$.

PROOF. The element $z = a^ib^j$ with $j \geq 0$ lies in the center if, and only if,

$$[z,b] = a^{i(1-r)} \quad \text{and} \quad [z,a] = a^{(r^j)-1}$$

are the unit element. This amounts to $t_o|i$ and $n_o|j$. The stated free presentation for $G/Z(G)$ is immediate. □

2.14 PROPOSITION. Let G be a metacyclic group and $\pi: G \to Q=G/Z(G)$ be the natural projection in e_G, cf. III.(1.1). Then $M(\pi): M(G) \to M(Q)$ vanishes and e_G is a generalized representation group of Q.

PROOF. From I.(3.3') we have an exact sequence

$$M(G) \xrightarrow{M(\pi)} M(Q) \twoheadrightarrow Z(G) \cap [G,G]$$

where the final arrow is the restriction of $\theta_*(e_G)$ onto its image. Let $G = G(m,n,r,\lambda)$, invoke the notation of the previous lemma. By this lemma $Q = G(t_o,n_o,r,0)$; the multiplicator of this group has order dividing

$$t_1 = (\frac{m}{(m,r-1)}, r-1)$$

by Theorem 2.10 and (2.4). On the other hand, $a^t \in Z(G) \cap [G,G]$ with

$$t = \mathrm{lcm}(t_o,(m,r-1)) = m/t_1 \quad ;$$

thus $Z(G) \cap [G,G]$ has order at least t_1 . We conclude that $\theta_*(e_G)$ is monomorphic and thus $M(\pi) = 0$. The final assertion follows from Proposition II.2.13. □

2.15 COROLLARY. Any two metacyclic groups with isomorphic central factor groups are isoclinic.

PROOF. Let G and H be metacyclic groups with isomorphic central factor groups. By Proposition 2.13 both $G \twoheadrightarrow G/Z(G)$ and $H \twoheadrightarrow H/Z(H) \cong G/Z(G)$ are generalized representation groups of $G/Z(G)$. Thus these extensions are isoclinic by Corollary III.2.4 (iii). Finally G and H are isoclinic groups, cf. Proposition III.1.4. □

2.16 COROLLARY. For fixed $m,n,r \in \mathbb{N}$ with $r^n \equiv 1 \mod m$, all groups $G(m,n,r,\lambda)$ lie in the same isoclinism family.

PROOF. Combine Corollary 2.15 with Lemma 2.13. □

3. The Precise Center of an Extension Group and Capable Groups

We recall that the idea of isoclinic groups was invented by P. HALL [1]. The concept of isoclinism of extensions, as introduced in Section III.1, basically has the advantage of allowing smoother formulations of the results. The added difficulties in the case of group isoclinism can be attributed to the bad behavior of the center with regard to homomorphisms $f: G \to H$, epimorphisms excepted, and to the fact that the Baer sum of two strictly central extensions is not strictly central in general. Here an extension

(3.1) $$e = (\varkappa,\pi) : A \rightarrowtail G \twoheadrightarrow Q$$

is, by definition, <u>strictly central</u> if $\varkappa(A) = Z(G)$. In order to utilize our previous results on isoclinism for the isoclinism classification of groups, we need and now give criteria for selecting the strictly central extensions from $\text{Cext}(Q,A)$. In the process we obtain various characterizations of capable groups, i.e. groups of the form $\text{Inaut}(G)$ for some G. The latter problem is interesting in its own right. The principal source for this section is BEYL/FELGNER/SCHMID [1], fundamental papers on the capability of abelian groups being due to BAER [2], [3].

3.1 DEFINITION (cf. HALL/SENIOR [1] and EVENS [1]). A group Q is <u>capable</u>, if it is isomorphic to $G/Z(G)$ for some group G, and <u>unicentral</u> if $\pi Z(G) = Z(Q)$ for each epimorphism $\pi: G \twoheadrightarrow Q$.

3.2 DEFINITION. For each generating subset J of the group G, let

$$W_J(G) = \bigcap_{x \in J} \langle x \rangle .$$

In other words, $x \in W_J(G)$ precisely when the roots of x contain J or $x = 1$. Define $W(G)$ as the join of all $W_J(G)$ where J runs through all generating subsets of G. Obviously $W(G)$ is a characteristic subgroup of G and

(3.2) $\qquad W(G)\cdot N/N \subseteq W(G/N) \qquad$ for $N \trianglelefteq G$.

3.3 PROPOSITION (cf. G.A. MILLER [1; p.339] or P. HALL [1; p.137]). Let e as in (3.1) be any central extension by Q. Then $W(Q) \subseteq \pi Z(G) \subseteq Z(Q)$. Thus a group Q with $W(Q) \neq 0$ is not capable.

PROOF. Let J be a generating subset of Q and $x \in J$. As e is central, $E_x := \pi^{-1}(\langle x\rangle)$ is abelian. Clearly

$$E_J := \pi^{-1}W_J(Q) = \bigcap_{x \in J} E_x ,$$

while G is the join of the E_x. It follows $E_J \subseteq Z(G)$ for all J. Finally $W_J(Q) = \pi(E_J) \subseteq \pi Z(G)$. □

3.4 EXAMPLES. When applicable, Proposition 3.3 is quite handy. However, the condition $W(Q) = 0$ does not imply the capability of Q.

a) Let A be a finite abelian group,

$$A \cong Z/n_1 \times \ldots \times Z/n_r \qquad \text{with } n_r|\ldots|n_1 ,$$

t_i being a generator of the i-th factor, of order n_i. Then one finds $W(A) = \langle t_1^{n_2}\rangle$ of order n_1/n_2. (Read $n_2 = 1$ in case $r = 1$.) Sketch of proof: Do the cases $r \leq 2$ by hand; the formula $\langle t_1^{n_2}\rangle \subseteq W(A)$ results from

$$J = \{ t_1, t_2\cdot t_1^{-1}, \ldots ,t_r\cdot t_1^{-1} \} ;$$

when $r \geq 3$, we can derive the reverse inclusion from the case $r = 2$ by invoking (3.2) for sufficiently many projections. Thus A is incapable for $n_1 \neq n_2$. We will see in Example 4.11 that A is capable when $n_1 = n_2$.

b) Let Q_8 be the quaternion group with generators i and j , let $\mathbb{Z}/2$ be generated by t . Then $G = Q_8 \times \mathbb{Z}/2$ is an incapable group with $Z(G) \cong G/Z(G) \cong \mathbb{Z}/2 \times \mathbb{Z}/2$ capable. This follows from $(-1,t^2) \in W(G)$, where $-1 := i^2 = j^2$. Indeed $(-1,t^2) \in W_J(G)$ where $J = \{(i,t^2),(j,t^2),(i,t)\}$ generates G .

c) Let p be an odd prime and G the extraspecial p-group of order p^5 and exponent p , cf. HUPPERT [1; p.355] or WARFIELD [1; Example 5.13]. It will be shown in 4.16 that G is unicentral, thus not capable. We are going to compute $W(G) = 0$. It suffices to consider minimal generating sets, say $J = \{x_1,\dots,x_4\}$. Assume we had

$$1 \neq z = x_1^m = x_2^n$$

for $m,n \in \mathbb{N}$. Since $x_1^p = x_2^p = 1$, it would follow that p doesn't divide m and $x_1 \in \langle x_2 \rangle$. This contradiction proves $W_J(G) = 0$.

3.5 PROPOSITION. If $\{N_i\}_{i\in I}$ is a system of normal subgroups of G such that all G/N_i are capable, then $G/(\bigcap_{i\in I} N_i)$ is also capable. Consequently G admits a least normal subgroup N with the property that G/N is capable.

PROOF. Consider the "diagonal" map

$$\Delta = \{g \longmapsto \{gN_i\}\} : G \to \prod_{i\in I} (G/N_i) .$$

Since $\operatorname{Ker} \Delta = \bigcap_{i\in I} N_i$, we have $\Delta(G) \cong G/(\bigcap_{i\in I} N_i)$ and wish to show that $\Delta(G)$ is capable. By assumption, there are strictly central extensions

$$e_i = (\varkappa_i,\pi_i) : A_i \rightarrowtail E_i \twoheadrightarrow G/N_i .$$

Then $\prod_{i\in I} e_i$ as in I.(2.3), but with index set I , is again strictly central. The restriction $e = \prod_{i\in I} e_i | \Delta(G)$ is a central extension of ΠA_i by $\Delta(G)$. It is easily checked to be strictly central and yields the capability of $\Delta(G)$. □

3.6 THEOREM. Let $e=(\varkappa,\pi): A \rightarrowtail G \twoheadrightarrow Q$ be a central extension and $\emptyset$ its "commutator form" as in Remark I.4.9. For the typical element $\bar{q} = q[Q,Q]$ of Q_{ab} and $z \in Z(Q)$,

(3.3) $\qquad \theta_*(e)\, \chi_Q(\bar{q} \otimes z) = \emptyset(q,z) \in A$

holds and

(3.4) $\qquad \pi Z(G) = \{ x \in Z(Q) \mid \forall\ \bar{q} \in Q_{ab} : \chi_Q(\bar{q} \otimes x) \in \mathrm{Ker}\ \theta_*(e) \}$.

Note that the final conclusion depends only on the subgroup

(3.5) $\qquad U(e) := \mathrm{Ker}\ \theta_*(e) \subseteq M(Q)$

rather than e itself - the same was true for isoclinism in Theorem III.2.3. In particular, $U(e)$ must be "small" for strictly central extensions e .

PROOF. Recall that $\emptyset(\pi g, \pi h) = \varkappa^{-1}[g,h]$ is defined whenever $x = \pi(g)$ and $y = \pi(h)$ commute in Q . Hence $\emptyset(y,z)$ is defined for all $y \in Q$ and all $z \in Z(Q)$. In view of $\pi Z(G) \subseteq Z(Q)$ we find

$$\pi Z(G) = \{ z \in Z(Q) \mid \forall\ y \in Q : \emptyset(y,z) = 1 \in A \} .$$

Thus (3.4) is reduced to the formula (3.3) which we now prove. Let $R \lhd F \twoheadrightarrow Q$ be the free presentation of Q at which $M(Q)$ shall be evaluated. From I.3.6 and $[\varkappa A,G] = 0$ we obtain the commutative diagram :

$$\begin{array}{ccccccc}
 & & & & F & \twoheadrightarrow & Q \\
 & & & & \downarrow{\scriptstyle \nu} & & \| \\
M(Q) & \hookrightarrow & \tilde{R} & \hookrightarrow & \tilde{F} & \twoheadrightarrow & Q \\
 & \searrow{\scriptstyle \theta_*(e)} & \downarrow & & \downarrow{\scriptstyle \beta} & & \| \\
 & & A & \overset{\varkappa}{\rightarrowtail} & G & \overset{\pi}{\twoheadrightarrow} & Q
\end{array}$$

where $\tilde{F} = F/[R,F]$ and $\tilde{R} = R/[R,F]$ and $\nu: F \to \tilde{F}$ is the natural projection. Pick $g,v \in G$ with $\pi(g) = q$ and $\pi(v) = z$ and then $f,w \in F$ with $\beta\nu(f) = g$ and $\beta\nu(w) = v$. Recall that χ_Q is the Ganea map of $e_Q: Z(Q) \hookrightarrow Q \twoheadrightarrow Q/Z(Q)$. Applying I.(4.3), with the groups Q and $Q/Z(Q)$ rather than G and Q, we obtain $\chi_Q(\bar{q} \otimes z) = \nu[f,w]$ and thus

$$\varkappa\, \theta_*(e)\, \chi_Q(\bar{q} \otimes z) = \beta\nu[f,w] = [g,v] = \varkappa\, \emptyset(q,z)\ .$$

As $\varkappa$ is monomorphic, we conclude (3.3). □

Since $0 = \{1\}$ is the smallest subgroup of $M(Q)$ and $U(e) = 0$ characterizes the generalized representation groups of Q, see Proposition II.2.13, the preceding theorem draws our attention to the characteristic subgroup

(3.6) $$Z^*(Q) = \{\, x \in Z(Q) \mid \forall\, q \in Q_{ab} : \chi_Q(q \otimes x) = 1 \,\}\ .$$

3.7 COROLLARY: Characterizations of $Z^*(Q)$.

(a) $Z^*(Q) = \mathrm{Ker}(\hat{\chi}_Q: Z(Q) \to \mathrm{Hom}(Q_{ab}, M(Q)))$,

where $\hat{\chi}_Q$ denotes the adjoint homomorphism of χ_Q.

(b) $Z^*(Q)$ is the image of the center of any generalized representation group, e.g. of a centralized free presentation of Q.

(c) $Z^*(Q) = \bigcap_e \pi Z(G)$,

where e ranges over all central extensions by Q as in (3.1). In particular, $Z^*(Q)$ contains $W(Q)$.

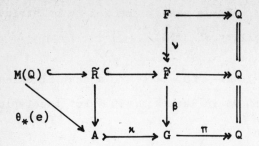

PROOF. Statement (a) reformulates the definition of $Z^*(Q)$. Now combine Theorem 3.6 with Proposition II.2.13 to obtain the assertions (b) and (c). The final remark follows from 3.3. □

In the case of finite groups Q , READ [1] already arrived at $Z^*(Q)$ via representation groups.

3.8 COROLLARY. For a group Q , the following are equivalent:

(i) Q is unicentral,

(ii) $Z^*(Q) = Z(Q)$,

(iii) $\chi_Q = 0 : Q_{ab} \otimes Z(Q) \to M(Q)$. □

Consequently perfect groups and groups with trivial multiplicator are unicentral. Combining Corollary 3.8 with the Ganea sequence of Theorem I.4.4, we find the following formula for the Schur multiplicator of a unicentral group Q :

$$(3.7) \quad \begin{aligned} M(Q) &\cong \mathrm{Ker}(\theta_*(e_Q): M(Q/Z(Q)) \to Z(Q)) \\ &= \mathrm{Ker}(M(Q/Z(Q)) \twoheadrightarrow Z(Q) \cap [Q,Q]) . \end{aligned}$$

3.9 PROPOSITION. (a) A group Q is capable precisely when $Z^*(Q) = 0$. A capable group is (isomorphic to) the center factor group of any generalized representation group of Q .

(b) $Z^*(Q)$ is the least normal subgroup N of Q such that Q/N is capable, cf. Proposition 3.5. In particular, $Q/Z^*(Q)$ is capable.

(c) Let M be a normal subgroup of Q outside $Z^*(Q)$, i.e. with $M \cap Z^*(Q) = 0$. If Q/M is capable, then so is Q .

PROOF. Part (a) is immediate from Corollary 3.7(b),(c); the second half was already proved as Proposition III.2.9. Concerning (b), we note that $Q/Z^*(Q) \cong G/Z(G)$ for any generalized representation group G of Q, by Corollary 3.7(b); thus $Q/Z^*(Q)$ is capable. Now let $N \trianglelefteq Q$ be such that Q/N is capable, $Q/N \cong E/Z(E)$ say. Then there is a commutative diagram

$$\begin{array}{lcccccc} e_\gamma : & Z(E) & \rightarrowtail & E^\gamma & \xrightarrow{\psi} & Q \\ & \| & & \downarrow{\scriptstyle \gamma^{\bullet}} & & \downarrow{\scriptstyle \gamma} \\ e : & Z(E) & \hookrightarrow & E & \xrightarrow{\pi} & Q/N \end{array}$$

with exact rows and $\operatorname{Ker} \gamma = N$. Since $\gamma^{\bullet}$ is surjective, we find $\gamma\psi Z(E^\gamma) = 0$ and thus $Z^*(Q) \subseteq \psi Z(E^\gamma) \subseteq N$. Finally, assume the hypotheses of (c). Then $Z^*(Q) = M \cap Z^*(Q) = 0$ by (b), thus Q is capable by (a). □

3.10 PROPOSITION. a) Let N be a central subgroup of Q, let $\nu: Q \twoheadrightarrow Q/N$ denote the natural projection. Then $N \subseteq Z^*(Q)$ precisely when $M(\nu): M(Q) \to M(Q/N)$ is monomorphic.

b) If $N \subseteq Z^*(Q)$ and i denotes the inclusion map into Q, then $\mathrm{res}=M(i): M(N) \to M(Q)$ vanishes.

PROOF. a) We invoke the exact sequence I.(4.2) for the central extension $e=(i,\nu): N \hookrightarrow Q \twoheadrightarrow Q/N$, viz.

$$Q_{ab} \otimes N \xrightarrow{\chi(e)} M(Q) \xrightarrow{M(\nu)} M(Q/N) .$$

Thus $M(\nu)$ is injective precisely when $\chi(e) = 0$. Now the definition of $\chi(e)$ in Theorem I.4.4(ii) exhibits $\chi(e) = 0$ as equivalent to $\chi_Q(x \otimes n) = 1$ for all $x \in Q_{ab}$ and all $n \in N$, in other words, to $N \subseteq Z^*(Q)$.

b) Since $M(0) = 0$, the composition $M(N) \to M(Q) \to M(Q/N)$ vanishes. Thus b) follows from a). □

The preceding results yield many easy-to-check conditions which imply that a group is not capable. We can only list a few of these.

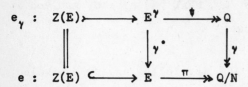

3.11 PROPOSITION. Assume G is a capable group and p a prime.

a) If $\mathrm{Hom}(G_{ab},M(G)) = 0$, then $Z(G) = 0$.

b) If one of G_{ab} or $M(G)$ has finite exponent n , then $Z(G)$ is bounded and its exponent divides n .

c) If G_{ab} is torsion, then $\mathrm{Tor}_p Z(G) \neq 0$ implies $\mathrm{Tor}_p G_{ab} \neq 0$.

d) If $M(G)$ is torsion, then $\mathrm{Tor}_p Z(G) \neq 0$ implies $\mathrm{Tor}_p M(G) \neq 0$.

e) If G_{ab} is torsion and there exists $x \in Z(G)$ of order p^k , for $k \geq 1$, then the Sylow p-subgroup S of $G_{ab}/\langle x[G,G]\rangle$ does not have exponent p^{k-1} .

Here $\mathrm{Tor}_p A$, for A abelian, denotes the subgroup of the elements of p-power order, i.e. the (unique) Sylow p-subgroup of A .

PROOF. Both (a) and (b) follow from the observation that $Z(G)$ is isomorphic to a subgroup of $\mathrm{Hom}(G_{ab},M(G))$, by Proposition 3.9(a) together with Corollary 3.7(a). If T is any torsion abelian group, then

$$\mathrm{Tor}_p T = 0 \quad \text{iff} \quad p^* \text{ is isomorphic,}$$

where $p^*(x) = x^p$ as in Remark 2.4. Therefore $\mathrm{Tor}_p T = 0$ implies $\mathrm{Tor}_p F(T) = 0$ for any additive functor F from the category of abelian groups to itself (terminology as in Remark 2.4). Concerning (c), if G_{ab} is torsion with $\mathrm{Tor}_p G_{ab} = 0$, then

$$\mathrm{Tor}_p Z(G) \subseteq \mathrm{Tor}_p \mathrm{Hom}(G_{ab},M(G)) = 0$$

by the preceding arguments. The proof of (d) is analogous. Finally, let us assume the hypotheses of (e) together with $q^*(S) = 0$, where $q = p^{k-1}$. We are going to derive $x^q \in Z^*(G)$, a contradiction to $Z^*(G) = 0$. Now $G_{ab} = P \times Q$ where P is the Sylow p-subgroup and

Q the Hall p'-subgroup of G_{ab} . Clearly $\chi_G(y \otimes x) = 0$ for all $y \in Q$ by order reasons and $\chi_G(x^m \otimes x) = 0$ by the formula I.(4.3). Therefore

$$\varphi = \{y \longmapsto \chi_G(y \otimes x)\} : G_{ab} \to M(G)$$

factors over $G_{ab} \twoheadrightarrow P \twoheadrightarrow S$. This implies $q^* \circ \varphi = 0$ and thus $\chi_G(G_{ab} \otimes x^q) = 0$ by the bilinearity of χ_G . □

3.12 REMARKS. a) A finite group Q that possesses a faithful irreducible complex projective representation (FICPR), must be capable: lift the FICPR to a linear representation of a representation group of Q and apply Schur's Lemma. The converse rarely holds, metacyclic groups constitute one of the exceptions - see Corollary 4.20.

b) The result of PAHLINGS [1] quoted in Remark II.3.13 implies the following: A finite group Q admits a FICPR precisely when there is a group G such that $G/Z(G) \cong Q$ and the socle (i.e. the join of the minimal normal subgroups) of G is generated by a single conjugacy class of G .

3.13 PROPOSITION. If Q is a finite capable group, then $\exp(Q)\cdot\exp(Z(Q))$ divides the order of Q .

This result is based on PAHLINGS [1; Satz 4.9].

PROOF. Let $e=(\varkappa,\pi): A \rightarrowtail G \twoheadrightarrow Q$ be a representation group of Q , then $Z(G) = \varkappa A \cong M(Q)$ by Propositions 3.9(a) and II.2.14. Since $\varkappa(A) = Z(G) \cap [G,G]$, we conclude that $\exp(\varkappa A)\cdot\exp(Q)$ divides $|Q|$ by a result of ALPERIN/KUO [1], cf. also BRANDIS [1]. Finally the exponent of $Z(Q)$ divides that of $M(Q)$ by Proposition 3.11(b). □

4. Examples of the Computation of $Z^*(G)$

Recall from Proposition 3.9 and Corollary 3.7(c) that the abelian groups $Z^*(G)$ and $Z(G)/Z^*(G)$ measure how much G deviates from being capable resp. unicentral. One may ask whether a particular group construction preserves capability or unicentrality. In general, a positive answer will require certain additional hypotheses, a typical verification being based on the computation of $Z^*(G)$. In the course we obtain $Z^*(G)$ for many specific groups G (extraspecial p-groups, finite metacyclic groups, various types of abelian groups). Most of the results are taken from BEYL/FELGNER/SCHMID [1].

4.1 PROPOSITION. Let G be a subdirect product of the groups $\{G_i\}_{i\in I}$. If all G_i are capable, then G is capable.

By definition, G is a subgroup of the (unrestricted) direct product $\prod_{i\in I} G_i$ such that $p_i(G) = G_i$ for all the natural projections p_i.

PROOF. Let E_i be groups with $E_i/Z(E_i) \cong G_i$. Then we obtain a morphism of extensions

$$\begin{array}{ccccc} A & \hookrightarrow & E & \twoheadrightarrow & G \\ \| & & \downarrow & & \downarrow \\ \prod_{i\in I} Z(E_i) & \hookrightarrow & \prod_{i\in I} E_i & \xrightarrow{\pi} & \prod_{i\in I} G_i \end{array}$$

where A is the center of $\prod E_i$ and E is defined as $\pi^{-1}(G)$. Since G is a subdirect product and $E \supseteq \operatorname{Ker} \pi$, the group E is a subdirect product of $\{E_i\}_{i\in I}$. Hence $\{t_i\} \in Z(E)$ implies $[t_i, E_i] = 0$ or $t_i \in Z(E_i)$ for all $i \in I$. This means $Z(E) = A$. □

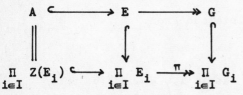

The preceding proposition implies

$$(4.1) \qquad Z^*(\underset{i\in I}{\times} G_i) \subseteq \underset{i\in I}{\times} Z^*(G_i)$$

for the restricted direct product $\underset{i\in I}{\times} G_i$, while equality does not hold in general. The minimal counterexample is given by $I = \{1,2\}$ and $G_1 = G_2 = \mathbb{Z}/2$. (Note that $\mathbb{Z}/2$ is unicentral while $\mathbb{Z}/2 \times \mathbb{Z}/2$ is capable.) The following result gives the precise conditions.

4.2 THEOREM. Assume $G = \underset{i\in I}{\times} G_i$ and let

$$\nu_i : Z^*(G_i) \hookrightarrow G_i \xrightarrow{ab} (G_i)_{ab}$$

be the obvious homomorphisms. Then $Z^*(G)$ consists precisely of those $z = \{z_i\}_{i\in I} \in \underset{i\in I}{\times} Z^*(G_i)$ that satisfy

$$(4.2) \qquad \bar{g}_i \otimes \nu_j(z_j) = 1 \in G_i \otimes G_j = (G_i)_{ab} \otimes (G_j)_{ab}$$

for all $i \neq j$ and all $\bar{g}_i \in (G_i)_{ab}$. Consequently $Z^*(G)$ coincides with the subgroup $\underset{i\in I}{\times} Z^*(G_i)$ of G , if and only if the homomorphisms

$$(4.3) \qquad 1 \otimes \nu_j : G_i \otimes Z^*(G_j) \to G_i \otimes (G_j)_{ab}$$

are trivial for all $i \neq j \in I$.

PROOF. a) Choose a representation group

$$e : A \overset{\varkappa}{\rightarrowtail} H \overset{\pi}{\twoheadrightarrow} G$$

of G with commutator from $\emptyset$ as in I.4.9 and 3.6, put $H_i = \pi^{-1}(G_i)$. We invoke $Z^*(G) = \pi Z(H)$ form Corollary 3.7(b) and note that $\pi^{-1}(z) \in Z(H)$ precisely when $\emptyset(g_i,z) = 1$ for all $g_i \in G_i$, $i \in I$. Since $[G_i,G_j] = 0$ for $i \neq j$, we have $[H_i,H_j]$ central in G . Hence the commutator form $\emptyset$ restricts to well-defined maps $\emptyset_{ij}: G_i \times G_j \to A$ which by Lemma I.4.1 are bihomomorphic, thus yield homomorphisms $\tilde{\emptyset}_{ij}: G_i \otimes G_j \to A$ with

$$\tilde{\emptyset}_{ij}(\pi(h_i)[G_i,G_i] \otimes \pi(h_j)[G_j,G_j]) = \kappa^{-1}([h_i,h_j])$$

for $h_i \in H_i$ and $h_j \in H_j$, whenever $i \neq j$. Write $\bar{g}_j = g_j[G_j,G_j]$.

b) Assume that z satisfies the conditions (4.2), to be shown $z \in Z^*(G)$. Fix i for the moment and let $g_i \in G_i$. Choose $h_i \in H_i$ and $t_j \in H_j$ with $\pi(h_i) = g_i$ and $\pi(t_j) = z_j$, $t_j = 1$ for almost all $j \in I$. As $e|G_i$ is a central extension and $z_i \in Z^*(G_i)$, it follows $[h_i,t_i] = 1$. Again by $[H_i,H_j] \subseteq Z(H)$, Lemma I.4.1(b) exhibits $\emptyset(g_i,z)$ as a product of elements $\emptyset(g_i,z_j)$ for finitely many $j \neq i$. Since

$$\emptyset(g_i,z_j) = \tilde{\emptyset}_{ij}(\bar{g}_i \otimes z_j) = \tilde{\emptyset}_{ij}(\bar{g}_i \otimes \nu_j z_j) = 1$$

for $i \neq j$ by (4.2), we conclude $\emptyset(g_i,z) = 1$, thus $z \in Z^*(G)$.

c) Assume $1_i \otimes \nu_j = 0$ for all $i \neq j$. Then all of $\times Z^*(G_j)$ satisfies (4.2), thus equality holds in (4.1).

d) Given $z=\{z_i\} \in Z^*(G)$, we are going to prove (4.2) for each pair $i,j \in I$ with $i \neq j$. To this end, let $Q = G_i \times G_j$ and $p: G \to Q$ be the natural projection. Formula (4.1) for the decomposition

$$G = Q \times \underset{k \neq i,j}{\times} G_k$$

yields $pZ^*(G) \subseteq Z^*(Q)$, thus $p(z) = (z_i,z_j) \in Z^*(Q)$. Now choose representation groups e_i and e_j of G_i and G_j , respectively; invoke Theorem II.4.11(c) to obtain a representation group $\bar{e}$ of Q. Let $\emptyset$ be the commutator form of $\bar{e}$. Again, as $\bar{e}|G_i$ is central, $z_i \in Z^*(G_i)$ implies $\emptyset(g_i,z_i) = 1$, thus $\emptyset(g_i,p(z)) = 1$ reduces to $\emptyset(g_i,z_j) = 1$. As in step (a), $\emptyset$ yields a homomorphism

$$\tilde{\emptyset} : G_i \otimes G_j \to \mathrm{Kernel}(\bar{e}) \cong M(Q)$$

with $\tilde{\emptyset}(\bar{g}_i \otimes \bar{g}_j) = \emptyset(g_i,g_j)$. This description of $\tilde{\emptyset}$ is seen to agree with the definition of the monomorphism $\kappa' \circ (\pi_1 \otimes \pi_2)^{-1}$ of

Theorem II.4.11(d). Consequently $\tilde{\emptyset}$ is monomorphic, and $(1 \otimes \nu_j)(\bar{g}_i \otimes z_j) = 1$ follows from

$$\tilde{\emptyset}\cdot(1 \otimes \nu_j)(\bar{g}_i \otimes z_j) = \tilde{\emptyset}(\bar{g}_i \otimes z_j) = \emptyset(g_i, z_j) = 1 .$$

e) Finally, if equality holds in (4.1), then we must have $\bar{g}_i \otimes \nu_j z_j = 1 \in G_i \otimes G_j$ for all $i \neq j$, by the previous step. This means the vanishing of all maps $1_i \otimes \nu_j$ for $i \neq j$. □

4.3 EXAMPLES. Conditions (4.3) are trivially fulfilled if all groups G_i are <u>stem groups</u>, i.e. satisfy $Z(G_i) \subseteq [G_i, G_i]$. They are also fulfilled if G is a nilpotent torsion group, $I = \{\text{primes } p\}$, and G_p is the Sylow p-subgroup of G. (Thus we assume that every element of G has finite order. If

$$G_p = \{ g \in G \mid g \text{ has p-power order } \}$$

is taken as a definition, then G_p is a normal subgroup and $G = \times G_p$. See e.g. WARFIELD [1; p.19] for these results of Baer and Hirsch.) The point is that $A \otimes B = 0$ whenever A is a (possibly infinite) p-group and B a q-group for primes $p \neq q$. In particular, all finite nilpotent and all torsion abelian groups G are included. Consequently, a nilpotent torsion group G is capable resp. unicentral precisely when so is each G_p.

4.4 PROPOSITION. Let U be a capable subgroup of finite index in G such that the transfer (Verlagerung) $\tau\colon G_{ab} \to U_{ab}$ is surjective. Then $Z^*(G) \cap U = 0$.

PROOF. Let $e \in \mathrm{Cext}(U,A)$ be a generalized representation group of U with factor system f, define $e'=\mathrm{Cor}^2(e) \in \mathrm{Cext}(G,A)$. Let $\{x_1,\ldots,x_m\}$ be a right transversal of U in G and $g_k = x_k g \cdot \overline{x_k g}^{-1} \in U$ in the notation of Corollary I.6.12. Then

$$\tau(y[G,G]) = \prod_{k=1}^{m} y_k[U,U]$$

for $y \in G$. For $g \in U \cap Z(G) \subseteq Z(U)$ we have $\overline{yg} = \overline{y}$ and $\overline{x_k g} = x_k$. Choosing the factor system f' of e' as suggested by Corollary I.6.12(b), we obtain

$$f'(y,g) = \prod_{k=1}^{m} f(y_k,g) \quad \text{and} \quad f'(g,y) = \prod_{k=1}^{m} f(g,y_k) .$$

We next recall from Remark I.5.9(b) the formula

$$\theta_*(e')\ \chi_G(y[G,G] \otimes g) = f'(y,g) \cdot f'(g,y)^{-1}$$

and a similar formula for e . Since χ_U is bihomomorphic,

$$\begin{aligned} \theta_*(e')\ \chi_G(y[G,G] \otimes g) &= \prod_{k=1}^{m} f(y_k,g) \cdot f(g,y_k)^{-1} \\ &= \theta_*(e)\ \chi_U\Big(\prod_{k=1}^{m} y_k[U,U] \otimes g\Big) \\ &= \theta_*(e)\ \chi_U(\tau(y[G,G]) \otimes g) . \end{aligned}$$

Now let even $g \in U \cap Z^*(G)$. Then $\chi_G(y[G,G] \otimes g) = 1$ for all $y \in G$, while $\theta_*(e)$ is monomorphic by Proposition II.2.13. Consequently

$$\chi_U(\tau(y[G,G]) \otimes g) = 1 \quad \text{for all} \quad y \in G .$$

Since τ is epimorphic by assumption, $g \in Z^*(U) = 0$. □

4.5 COROLLARY. Let N be a normal subgroup of G of finite index such that the transfer $G_{ab} \to N_{ab}$ is surjective. If both N and G/N are capable, then G is capable.

PROOF. Proposition 4.4 gives $Z^*(G) \cap N = 0$. Apply Proposition 3.9(c). □

4.6 EXAMPLE. Let $N \lhd G \twoheadrightarrow Q$ be a group extension such that Q is capable and N is centerless perfect. Then N is clearly capable and the transfer is trivially surjective. For an example involving symmetric and alternating groups, let $G = S_n \times \mathbb{Z}/2$ and $N = A_n \times 0$ for $n \geq 5$, thus $S_n \times \mathbb{Z}/2$ is capable. Actually

$Z^*(S_n \times \mathbb{Z}/2) = 0$ holds already for $n \geq 2$ by Theorem 4.2.

The Ganea map of an abelian group S is $\chi(A): A \otimes A \to M(A)$ and induces a natural isomorphism $\chi_o: A \wedge A \to M(A)$ by Theorem I.4.7. The connection is

(4.4) $\qquad \chi_A(a \otimes b) = \chi_o(a \wedge b) \in M(A) \quad$ for all $a,b \in A$.

Combining this with (3.6), we obtain

4.7 PROPOSITION. If A is an abelian group, then

$$Z^*(A) = \{ x \in A \mid \forall a \in A : a \wedge x = 0 \in A \wedge A \} . \quad \square$$

4.8 EXAMPLE. If the abelian group A is divisible torsion or locally cyclic, then clearly $M(A) \cong A \wedge A \cong 0$ and thus $Z^*(A) = A$. Actually, comparing the definitions in 3.1 and I.3.10(c), we see that an abelian group A is unicentral precisely when it is absolutely abelian. Now Corollary 3.8 together with I.4.5 gives a different proof of the fact that the latter is equivalent to $M(A) = 0$. This topic is resumed in Example 6.17(b).

4.9 PROPOSITION. Let $G = \times_{i \in I} (\mathbb{Z}/n_i)$ be a direct sum (restricted direct product) of cyclic groups ($n_i = 0$ allowed). Let T denote the torsion subgroup and $r = \dim_{\mathbb{Q}}(G \otimes \mathbb{Q})$ the rational rank of G . Then

a) $Z^*(G) = 0$ for $r \geq 2$.

b) Suppose $r = 1$. If T is unbounded, then $Z^*(G) = 0$. If T is of finite exponent m , then $Z^*(G) = G^m = \{ g^m \mid g \in G \}$ is cyclic; a generator is x^m for any infinite direct summand $\{x\}$ of G .

c) Suppose $r = 0$, i.e. $G = T$ torsion. Let P be the set of primes for which the Sylow p-subgroup G_p is bounded. For each

$p \in P$ let $n(p)$ the least power of p such that $(G_p)^{n(p)}$ is cyclic. (In other words, $n(p)$ is the second "torsion coefficient" from the top of G_p .) Then

$$(4.5) \qquad Z^*(G) = \underset{p \in P}{\times} (G_p)^{n(p)} .$$

PROOF. By further decomposing, if necessary, we may assume that each n_i is 0 or a prime-power. We use additive notation for convenience. Let x_i be a generator of $\mathbb{Z}/n_i$. By Proposition 4.7, the typical element $z = \sum \lambda_i x_i$ lies in $Z^*(G)$ precisely when $\lambda_i x_i \wedge x_j = 0$ for all $j \in I$. For $i \neq j$, $x_i \wedge x_j$ is a generator of $\mathbb{Z}/(n_i,n_j)$, thus $\lambda_i x_i \wedge x_j = 0$ is equivalent to $(n_i,n_j) \mid \lambda_i$. The discussion of the various cases is now straightforward. In particular, whenever $n_j = 0$, then $z \wedge x_j = 0$ implies $n_i \mid \lambda_i$ for all $i \in I \setminus \{j\}$, i.e. $z \in \{xj\}$. □

4.10 COROLLARY (BAER [3]). Let G be a direct sum of cyclic groups as above. Then G is capable precisely when one of the following holds:

(i) the rational rank r of G exceeds 1 ; or

(ii) $r = 1$ and Tor(G) is unbounded; or

(iii) G is torsion and for each prime p , either the Sylow p-subgroup G_p is unbounded or the two highest torsion coefficients of G_p agree. □

4.11 EXAMPLE. Consider the special case that G is finitely generated abelian. Then Proposition 4.9(c) implies that $Z^*(G)$ is cyclic, indeed $Z^*(G) = W(G)$ as in 3.4(a). In particular, we obtain the well-known result that A is capable precisely when the two highest torsion coefficients agree.

4.12 REMARK. An explicit description of all capable abelian groups seems to be difficult, mainly because a directed union (direct limit) of capable groups need not be capable. A counterexample is the unicentral group $G = \mathbb{Z}(p^\infty) \times \mathbb{Z}(p^\infty)$ which is the directed union of the capable subgroups $\mathbb{Z}/p^k \times \mathbb{Z}/p^k$, $k \in \mathbb{N}$. A result of MOSKALENKO [1; §4] implies that an unbounded abelian p-group G is capable if, and only if, no non-zero element of G has infinite height. A bounded abelian p-group is a direct sum of finite cyclic groups, this case is covered by Corollary 4.10(iii).

4.13 PROPOSITION. Let A be a torsion-free abelian group. If $r = 1$ for the rational rank r of A, then A is unicentral and not capable. If $r \neq 1$, then A is capable.

PROOF. We use that A is the directed union of its finitely generated free subgroups. If $r = 1$, then A is locally cyclic and thus $M(A) = 0$. The case $r = 0$ means the trivial group. Now assume $r \geq 2$, write $A = \cup B_\alpha$ where B_α runs through all free-abelian subgroups of A of rank at least two. We claim that any inclusion $j: B_1 \hookrightarrow B_2$ induces a monomorphism $M(j): M(B_1) \to M(B_2)$. Indeed, the homomorphism j between free-abelian groups can be described by $j(x_i) = n_i y_i$ with $n_i \in \mathbb{Z}$ with respect to suitable bases $\{x_1, \ldots, x_m\}$ of B_1 and $\{y_1, \ldots, y_n\}$ of B_2. (Apply elementary row and column operations to the companion matrix of j, see SEIFERT/THRELFALL [1; §76] for an explicit procedure.) In the present situation, $2 \leq m \leq n$ and $n_i \neq 0$. In view of the natural isomorphism $M(B_i) \cong B_i \wedge B_i$, the claim is immediate.

Let $a \in Z^*(A)$, then there exists a free-abelian subgroup B of rank two with $a \in B$. Since $M(A) \cong \text{dir.lim.} M(B_\alpha)$ by the direct limit argument I.5.10, we conclude that $j: B \hookrightarrow A$ induces a monomorphism $M(B) \to M(A)$. Now the commutative diagram

$$\begin{array}{ccc} B \otimes B & \xrightarrow{\chi(B)} & M(B) \\ {\scriptstyle j \otimes j}\downarrow & & \downarrow{\scriptstyle M(j)} \\ A \otimes A & \xrightarrow{\chi(A)} & M(A) \end{array}$$

yields $M(j)\ \chi_B(a \otimes B) = \chi_A(a \otimes jB) = 0$, thus $\chi_B(a \otimes B) = 0$ and $b \in Z^*(B) = 0$. In the last step, we also used Proposition 4.9(a). □

4.14 PROPOSITION. Let A be an abelian group and U a pure subgroup of A , e.g. Tor(A). If U and A/U are both capable, then so is A .

By definition U is <u>pure</u> in A , if $U \cap A^n = U^n$ holds for all $n \in \mathbb{N}$.

PROOF. Let $\varkappa \colon U \to A$ be the embedding. Corollary 3.9(b) implies $Z^*(A) \subseteq U$. Let $a \in Z^*(A)$. As in the previous proof, we have

$$M(\varkappa)\ \chi_U(a \otimes U) = \chi_A(a \otimes \varkappa U) = 0 .$$

$$\begin{array}{ccc} U \otimes U & \xrightarrow{\chi(U)} & M(U) \\ {\scriptstyle \varkappa \otimes \varkappa}\downarrow & & \downarrow{\scriptstyle M(\varkappa)} \\ A \otimes A & \xrightarrow{\chi(A)} & M(A) \end{array}$$

We now invoke from BEYL [3; Thm. 1.6] that $M(\varkappa)$ embeds $M(U)$ into $M(A)$ as a pure subgroup. (The point is that $U \hookrightarrow A \twoheadrightarrow A/U$ is a direct limit of split extensions $U \hookrightarrow A_i \twoheadrightarrow B_i$. The splitting exhibits $M(U)$ as a direct summand of $M(A_i)$. Finally $M(\varkappa)$ is the direct limit of monomorphisms.) Consequently $\chi_U(a \otimes U) = 0$ and $a \in Z^*(U) = 0$. □

4.15 THEOREM. Given an abelian group A and a family of stem groups $\{G_i\}_{i \in I}$. Assume $|I| \geq 2$ and $\varphi_i \colon A \cong Z(G_i)$ for all $i \in I$. Then the central product G of this family is unicentral.

The <u>central product</u> G is that quotient of $X = \times_{i\in I} G_i$ in which precisely the centers $Z(G_i)$ are identified. More formally, the definition is $G = X/N$ with

$$N = \{ \{z_i\} \in \times_{i\in I} Z(G_i) \subseteq X \mid \prod_{i\in I} \varphi_i^{-1}(z_i) = 1 \in A \} .$$

PROOF. Let $\sigma: X \to X/N = G$ denote the natural projection. We first observe that each composite map

$$\nu_j : Z(G_j) \hookrightarrow G_j \hookrightarrow X \xrightarrow{\sigma} G$$

has image $Z(G)$. Choose any (generalized) representation group

$$e = (\varkappa, \pi) : M \rightarrowtail E \twoheadrightarrow G$$

of G; let $E_i = \pi^{-1}(G_i)$ for $i \in I$, let $\emptyset$ be the "commutator form" of e as in I.4.9. Given any $g_i \in G_i$, $z_j \in Z(G_j)$ for $i,j \in I$, we claim

(4.6) $\quad \emptyset(\sigma g_i, \nu_j z_j) = 1 \in M$.

If $i = j$, pick $k \in I$ with $k \neq i$ and use $\nu_j z_j = \nu_k z_k$ with $z_k = \varphi_k \varphi_i^{-1}(z_i)$. Thus we also assume $i \neq j$ until (4.6) is proved. Since $[G_i, G_j] = 0$, we have $[E_i, E_j] \subseteq \varkappa M$ central in E and $[E_i,[E_j,E_j]] = 0$ by the Three-Subgroups Lemma I.4.3. But $z_j \in [G_j, G_j]$ by the assumption on G_j, hence

$$\varkappa\emptyset(\sigma g_i, \nu_j z_j) \in [E_i,[E_j,E_j]] .$$

Since e is central and the elements σg_i generate the group G, (4.6) implies $\pi^{-1}(\nu_j z_j) \in Z(E)$, for each $z_j \in Z(G_j)$. From this and the initial remark, we conclude $\pi Z(E) = Z(G)$. Thus G is unicentral by Corollary 3.7(b). □

Let p be any prime number. An <u>extra-special</u> p-group is a group G with $Z(G) = [G,G] \cong \mathbb{Z}/p$ and G_{ab} of exponent p. For finite G one has $|G| = p^{2k+1}$ with $k \geq 1$.

4.16 COROLLARY: Multiplicators of extra-special p-groups.

a) For $2 \leq k \leq \infty$, every extra-special p-group G of order p^{2k+1} is unicentral and not capable. The Schur multiplicator of G is elementary abelian of p-rank $2k^2 - k - 1$.

b) The quaternion group Q_8 has trivial multiplicator, thus is unicentral. The dihedral group D_8 of order 8 is capable and has multiplicator $\mathbb{Z}/2$. For p odd, the extra-special p-group H of order p^3 and exponent p is capable and has multiplicator $\mathbb{Z}/p \times \mathbb{Z}/p$, whereas the extra-special p-group of order p^3 and exponent p^2 has trivial multiplicator, thus is unicentral.

Part (b) corrects errors in OPOLKA [1; Lemma (1.4)] and in BEYL/FELGNER/SCHMID [1; Proof 8.2 p.174], in the latter case read "stem extension" instead of "stem cover".

PROOF. a) Since $[G,G] \neq 0$, there are elements $x,y \in G$ with $xy \neq yx$. Let $G_1 = \langle x,y \rangle$ and G_2 the centralizer of G_1 in G . We claim that G_1 and G_2 are stem groups with $Z(G_1) = Z(G) = Z(G_2)$ and that G is (isomorphic to) the central product of G_1 and G_2 . This being granted, G is unicentral by Theorem 4.15. Now $M(G_{ab}) \cong G_{ab} \wedge G_{ab}$ is elementary abelian of rank $k(2k-1)$ by Theorem I.4.7. Finally (3.7) exhibits $M(G)$ as a subgroup of $M(G_{ab})$ with rank defect one.

We are left to prove the claim. Since G_1 is a non-abelian p-group of order at most p^3 , it must be extra-special of order p^3 . In particular, $Z(G_1) = [G_1,G_1] = [G,G] = Z(G)$. An arbitrary $g \in G$ may be written in the form $g = x^{\varkappa} \cdot y^{\lambda} \cdot z$ with $z \in G_2$, basically because $[x,y]$ generates $[G,G]$, has order p , and is central. (The first step is to select $\varkappa$ such that $x^{-\varkappa}g$ commutes with y .) We conclude $G = G_1 \cdot G_2$. This and the definition of G_2 imply

$G_1 \cap G_2 = Z(G)$. By order reasons G_2 is not contained in G_1 and $|G_2| > p$. Now $|Z(G)| = p$ implies $Z(G_2) = Z(G)$ rather than proper inclusion. Thus G_2 is not abelian and $[G_2,G_2] = [G,G]$. By the previous steps the multiplication map $G_1 \times G_2 \to G$ is surjective with kernel isomorphic to $Z(G)$, exhibiting G as the central product of G_1 and G_2 .

b) The multiplicators of Q_8 and D_8 have been obtained in Example II.3.8. The reader checks that D_8 is the central factor group of the dihedral group D_{16} . The assertions on H have been proved in Example 1.12. The remaining group G is the metacyclic group $G(p^2,p,1+p,0)$ and has trivial multiplicator by Theorem 2.10; or one can prove $W(G) = Z(G)$ directly. We mention that LEWIS [1] actually determined the integral cohomology of the non-abelian groups of order p^3 ; the multiplicators can also be computed from his results. □

4.17 REMARK. By the preceding corollary there is precisely one capable extra-special p-group for each prime p . These groups do possess a faithful irreducible complex projective representation (FICPR) , as was noted by NG [2; Prop. (7.6)] and OPOLKA [1; Lemma (1.2)]. Alternatively, the criterion of Pahlings (as stated in 3.12) is satisfied in each case. First, $D_8 \cong D_{16}/Z(D_{16})$ with $\mathrm{Socle}(D_{16}) = Z(D_{16}) \cong \mathbb{Z}/2$ cyclic. For p odd, the group H of order p^3 and exponent p is the central factor group of group No. 12, call it $\tilde{H}$, in HUPPERT [1; Satz III.12.6] with $\mathrm{Socle}(\tilde{H}) = Z(\tilde{H}) \cong \mathbb{Z}/p$. (Our assertion is also valid for $p = 3$, but then $\tilde{H}$ has exponent 9 rather than 3.)

4.18 PROPOSITION. Let $G = G(m,n,r,\lambda)$ be the metacyclic group defined in 2.3 with $r^n \equiv 1 \bmod m$ and $\lambda | h(m,n,r)$. Then $Z^*(G)$ is the cyclic group of order $\frac{n}{k} \cdot \frac{(m,r-1)}{\lambda}$ generated by b^k , where

k divides n and is the smallest positive integer ν such that $m\lambda/(m,r-1)$ divides $1 + r +\ldots+ r^{\nu-1}$.

PROOF. Let $\varphi: \tilde{G}=G(m\cdot\lambda,n,s,1) \to G$ be the representation group constructed in the proof of Theorem 2.10, the notation and assumptions of which are preserved. Then $Z^*(G) = \rho Z(\tilde{G})$ by Corollary 3.7. The center of $\tilde{G}$ is generated by $\tilde{a}^{m\lambda/(m,s-1)}$ and $\tilde{b}^l$ where l is the least positive integer with $s^l \equiv 1 \mod m\cdot\lambda$, see Lemma 2.13. Now $l|n$ by (2.6) and thus

$$\tilde{a}^{m\lambda/(m,s-1)} = \tilde{b}^n$$

is a power of $\tilde{b}^l$. Hence $Z^*(G)$ is the cyclic group generated by b^l . We recall that $\frac{s-1}{(m,r-1)} = t$ is prime to m . Then $k = l$ follows from

$$m\cdot\lambda \mid s^\nu-1 \quad \text{iff} \quad \frac{m}{(m,r-1)/\lambda} \mid t\cdot(1+s+ \ldots +s^{\nu-1})$$

$$\text{iff} \quad \frac{m\cdot\lambda}{(m,r-1)} \mid 1+r+ \ldots +r^{\nu-1} . \quad \square$$

4.19 COROLLARY. The metacyclic group $G = G(m,n,r,\lambda)$, with the parameters satisfying $r^n \equiv 1 \mod m$ and $\lambda \mid h(m,n,r)$, is capable if, and only if, $\lambda = (m,r-1)$ and n is the smallest positive integer such that m divides $1+r+ \ldots +r^{n-1}$.

The first numerical condition implies that $e(m,n,r,\lambda)$ as in 2.3 is the split extension. We see that then $m = |M(G)|\cdot|[G,G]|$ and $n = |G|/m$ are invariants of G .

PROOF. The condition is $Z^*(G) = 0$, which becomes $n = k$ and $\lambda = (m,r-1)$. This implies $m \mid 1+r+ \ldots +r^{n-1}$ and $h(m,n,r) =$ $= (m,r-1)$. For the converse, first note that the listed numerical conditions imply $r^n \equiv 1 \mod m$ and $\lambda = h(m,n,r)$. $\square$

4.20 COROLLARY. A metacyclic group has a faithful irreducible complex projective representation precisely when it is capable.

Actually NG [2; Theorem (4.4)] obtained the metacyclic groups with an FICPR without noting the connection with capability, his method was to construct an FICPR explicitly.

PROOF. Assume that $G = G(m,n,r,\lambda)$ is capable and again let $\rho: \tilde{G} \to G$ be the representation group (2.7). First $Z(\tilde{G}) = \operatorname{Ker} \rho$ by Proposition III.2.9 or Corollary 3.9. We claim that every minimal subgroup of $\tilde{G}$ lies in $\{\tilde{a}\}$; this means that the socle of $\tilde{G}$ is cyclic and the criterion of Pahlings applies (see 3.12). On the contrary, assume that there is a minimal normal subgroup N containing $z = \tilde{a}^i\tilde{b}^j$ with $0 < j < n$. Then

$$\tilde{a}\cdot z\cdot \tilde{a}^{-1}\cdot z^{-1} = \tilde{a}^{(1-s^j)} \in \{\tilde{a}\} \cap N$$

generates a normal subgroup of $\tilde{G}$ inside N , hence $s^j \equiv 1 \mod m\cdot\lambda$ by minimality. By Proposition 4.18 (cf. the last step of its proof) we conclude $b^j \in Z^*(G)$, contradiction. □

4.21 PROPOSITION. A metacyclic group G is unicentral precisely when $M(G) = 0$.

PROOF. Since $\chi(G): G_{ab} \otimes Z(G) \to M(G)$ is epimorphic by Proposition 2.14 and Theorem I.4.4(ii), $\chi(G)$ vanishes precisely when $M(G) = 0$. Invoke Corollary 3.8. □

5. Preliminaries on Group Varieties

While the basic reference for group varieties is the book by H. NEUMANN [1], our view is close to STAMMBACH's [3; chp. I]. In order to keep these notes self-contained, we here establish some notation and present a few facts of particular relevance to us.

Let F_∞ be the free group of countable rank on the generators $x_1, x_2, x_3, \ldots$. A set of laws is just a subset of F_∞ . If G is any group, the verbal subgroup VG of G is defined as the subgroup generated by

$$\{ f(v) \mid \text{all homomorphisms } f: F_\infty \to G \text{ , all } v \in V \} .$$

The group G is said to satisfy the laws V if $VG = 0$. The variety $\mathfrak{V}$ belonging to V is defined as the full subcategory of the category $\mathfrak{G}$ of "all" groups, the objects of which are the groups satisfying V . If $V = VF_\infty$ (in other words, if V is fully invariant), then V is called a closed set of laws. Every subset $W \subseteq F_\infty$ determines a least closed set containing it, viz. $V = WF_\infty$; then $WG = VG$ for each group G . For example, each of $V_1 = \{[x_1,x_2]\}$, $V_2 = \{[x_2,x_3{}^2],[x_5,x_7]\}$, and $V_3 = [F_\infty,F_\infty]$ defines the variety $\mathfrak{A}$ of "all" abelian groups. The fully invariant subgroups of F_∞ are in bijective correspondence with group varieties such that $V \subseteq W$ iff $\mathfrak{W} \subseteq \mathfrak{V}$, cf. H. NEUMANN [1; Thm. 14.31].

5.1 REMARK. Let $\mathfrak{V}$ be the variety belonging to the laws V . Whenever $h: G \to H$ is epimorphic, one has $VH = h(VG)$. Thus $V(G/VG) = 0$ and G/VG is the largest factor group of G in the variety $\mathfrak{V}$.

5.2 DEFINITION. Let $\mathfrak{V}$ be the variety belonging to the laws V. If $e: R \hookrightarrow F \twoheadrightarrow Q$ is a free presentation of Q as in I.3 and $Q \in \mathfrak{V}$, then $VF \subseteq R$ and

$$(5.1) \qquad e/VF : \frac{R}{VF} \hookrightarrow \frac{F}{VF} \xrightarrow{\tau} Q$$

is called a $\mathfrak{V}$-free presentation of Q.

Indeed, the groups F/VF lie in $\mathfrak{V}$ by 5.1 and are $\mathfrak{V}$-free, i.e. free in the category $\mathfrak{V}$ with respect to the functor "underlying set". Every $\mathfrak{V}$-free group is isomorphic to F/VF for a suitable free group F. See STAMMBACH [3; I §3] for the details of this aspect.

5.3 BIRKHOFF's THEOREM. Every non-empty full subcategory of $\mathfrak{G}$ that is closed under subgroups, epimorphic images (i.e. groups isomorphic to factor groups), and (unrestricted) direct products, is a group variety, i.e. defined by some set of laws. The converse is also true.

PROOF, cf. H. NEUMANN [1; Thm. 15.51]. Indeed the converse assertion is straightforward. Now let $\mathfrak{V}$ be a subcategory with closure properties as specified. For each $Q \in \mathfrak{V}$ and each homomorphism $f: F_\infty \to Q$, let $K_f = \operatorname{Ker} f \trianglelefteq F_\infty$ and

$$(5.2) \qquad V \underset{\text{def}}{=} \bigcap_{\text{all } f} K_f .$$

Since the totality of normal subgroups of F_∞ is a set, this intersection is well-defined, even if "the totality of all f" is not a set. If $G \in \mathfrak{V}$ and $f: F_\infty \to G$ is any homomorphism, then $\operatorname{Ker} f \supseteq V$; thus G satisfies the laws V.

If Q is any group satisfying the laws V, then it is an epimorphic image of F/VF for some free group F of infinite rank. Hence it suffices to prove $F/VF \in \mathfrak{V}$ for such F. Consider Ker f for all $G \in \mathfrak{B}$ and all homomorphisms $f: F \to G$. The totality I of these normal subgroups of F is a non-empty set, for each $i \in I$ select $G_i \in \mathfrak{B}$ and $f_i: F \to G_i$ such that $\operatorname{Ker} f_i = i$, put

$$W = \bigcap_{i \in I} \operatorname{Ker} f_i .$$

As G_i satisfies the laws V, we have $i \supseteq VF$. Since $\mathfrak{B}$ is subgroup and product closed, all $F/i \cong \operatorname{Im} f_i \subseteq G_i$ and then

$$P = \prod_{i \in I} F/\operatorname{Ker} f_i$$

lie in $\mathfrak{B}$. There is an obvious homomorphism $\varphi: F/VF \to P$ with

$$\operatorname{Ker} \varphi = \bigcap_{i \in I} \frac{\operatorname{Ker} f_i}{VF} = \frac{W}{VF} .$$

We claim $W = VF$, thus φ is monomorphic and $F/VF \in \mathfrak{B}$. To this end, fix a basis of F and observe that only finitely many basis elements occur in any single $x \in W$. One thus finds $\tilde{x} \in F_\infty$ and an embedding $\lambda: F_\infty \to F$ as a free factor with $\lambda(\tilde{x}) = x$. Every homomorphism $\tilde{f}: F_\infty \to G \in \mathfrak{B}$ can be extended to $f: F \to G$ such that $f \circ \lambda = \tilde{f}$. By the very definition of I and W, we have $\tilde{f}(\tilde{x}) = f(x) = 1$. We conclude $\tilde{x} \in V$ and then $x = \lambda(\tilde{x}) \in VF$. □

On the basis of the preceding, we feel free to call VG the $\mathfrak{B}$-verbal subgroup $\mathfrak{B}G$ of G and to use similar confusions below. Moreover, Birkhoff's Theorem elucidates how any class of groups generates a variety without explicit mention of laws. We typically name a variety after the groups in it, i.e. "the variety $\mathfrak{A}$ of abelian groups".

5.4 LEMMA. Let U and V be closed sets of laws and $W = [U,V]$. Then $WG = [UG,VG]$ for all groups G.

PROOF. Clearly $WG \subseteq [UG,VG]$. Recall that $[UG,VG]$ is the subgroup generated by all $[x,y]$ with $x \in UG$ and $y \in VG$. There suffice finitely many $u_i \in U$ and $f_i\colon F_\infty \to G$ to exhibit

$$x = \prod_{i=1}^{n} f_i(u_i)^{\pm 1} .$$

Since each u_i involves only finitely many of the generators $x_1, x_2, \ldots$ and U is closed, there exists $u \in U$ and $f\colon F_\infty \to G$ with $x = f(u)$ - idea: choose disjoint sets of generators for different u_i. Likewise, $y = g(v)$ for some $g\colon F_\infty \to G$ and $v \in V$ such that u and v have no generator x_i in common. Then $[x,y] = h([u,v])$ with $[u,v] \in W$ and a homomorphism h agreeing with f resp. g on the relevant generators. □

5.5 DEFINITION. Let $\mathfrak{W}$ be any variety, defined by the closed set of laws W. Then the groups G with $G/Z(G) \in \mathfrak{W}$ or, equivalently, $WG \subseteq Z(G)$ form a variety; this is called the variety of <u>center-by-$\mathfrak{W}$</u> groups. It follows from Lemma 5.4 that the variety of center-by-$\mathfrak{W}$ groups is defined by the laws $[F_\infty, W]$. For example, let $\mathfrak{N}_c$ denote the variety of nilpotent groups of class at most c. Then $\mathfrak{N}_1 = \mathfrak{A}$ and $\mathfrak{N}_{c+1}$ = center-by-$\mathfrak{N}_c$ for $c \geq 1$. On the other hand, the totality of groups G with WG abelian form a variety; equivalently, we deal with groups G supported by an extension $A \rightarrowtail G \twoheadrightarrow Q$ with A abelian and $Q \in \mathfrak{W}$. Therefore the variety is called the variety of <u>abelian-by-$\mathfrak{W}$</u> groups; it is defined by the laws $[W,W]$ according to Lemma 5.4. For example, let $\mathfrak{S}_l$ denote the variety of soluble groups of length at most l. Then $\mathfrak{S}_1 = \mathfrak{A}$ and $\mathfrak{S}_{l+1}$ = abelian-by-$\mathfrak{S}_l$ for $l \geq 1$.

5.6 REMARK. For every group variety $\mathfrak{B}$, the following are equivalent:

(i) The variety contains the integers $\mathbb{Z}$;

(ii) $\mathfrak{B}$ contains all abelian groups;

(iii) $\mathfrak{B}$ is defined by commutator laws, i.e. $V \subseteq [F_\infty, F_\infty]$.

Such a variety is said to have exponent zero. Otherwise, there exists an integer $q > 0$ such that $\mathbb{Z}/q$ but no larger cyclic group lies in $\mathfrak{V}$; then $\mathfrak{V}$ is said to have exponent q .

Actually, the implications (iii) ⇒ (ii) ⇒ (i) are trivial. If $\mathbb{Z} \in \mathfrak{V}$ and v is any law of $\mathfrak{V}$, then the exponent sum with respect to each x_i must vanish, hence $v \in [F_\infty, F_\infty]$. If $\mathfrak{V}$ contains cyclic groups $\mathbb{Z}/n_i$ of arbitrary large finite order, then $\mathfrak{V}$ also contains $P = \Pi\, \mathbb{Z}/n_i$ and $\mathbb{Z}$ (as a subgroup of P), hence has exponent zero. Finally, if $\mathfrak{V}$ has exponent q and $V = \mathfrak{V}F_\infty$, then V is generated by x_1^q and $V \cap [F_\infty, F_\infty]$, cf. H. NEUMANN [1; Thm. 12.12].

5.7 REMARK. Every variety $\mathfrak{V}$ determines a smallest variety of exponent 0 containing $\mathfrak{V}$. This variety is generated by the groups of $\mathfrak{V}$ together with $\mathbb{Z}$, we call it varo($\mathfrak{V}$) . It is defined by the laws $[F_\infty, F_\infty] \cap \mathfrak{V}F_\infty$.

5.8 DEFINITION. We say that an extension $e: N \rightarrowtail G \twoheadrightarrow Q$ lies in the variety $\mathfrak{V}$, if $G \in \mathfrak{V}$. If $e \equiv e_1$ and $e_1 \in \mathfrak{V}$, then certainly e lies in $\mathfrak{V}$, we write $e \in \mathfrak{V}$. For $Q \in \mathfrak{V}$ and $A \in \mathfrak{A} \cap \mathfrak{V}$, let

$$\mathrm{Cext}_{\mathfrak{V}}(Q,A) \subseteq \mathrm{Cext}(Q,A)$$

denote the set of classes of central extensions of A by Q that lie in $\mathfrak{V}$. It follows from the direct constructions of I.1 and the formula I.(2.2) that $\mathrm{Cext}_{\mathfrak{V}}(Q,A)$ is a subgroup of $\mathrm{Cext}(Q,A)$ and a functor $\mathfrak{V}^{op} \times (\mathfrak{V} \cap \mathfrak{A}) \to \mathfrak{A}$.

If $v = v(x_1, \ldots, x_n) \in F_\infty$ is a word in the first n basis elements (variables) and $g_1, \ldots, g_n$ are elements of the group G , then $v(g_1, \ldots, g_n)$ designates the element $f(v) \in G$ where

$f\colon F_\infty \to G$ is the homomorphism with $f(x_i) = g_i$ for $i \leq n$ and $f(x_j) = 1$ for $j > n$. If $\alpha\colon G \to H$ is any homomorphism, there results the formula

$$(5.3) \qquad v(\alpha g_1,\ldots,\alpha g_n) = \alpha v(g_1,\ldots,g_n)\ .$$

6. Central Extensions and Varieties

The basic theory on the problem whether the middle group of a group extension lies in a given variety $\mathfrak{B}$, can be found in STAMMBACH [3]. In this section we obtain rather explicit (computable) criteria for this question, provided $\mathfrak{B}$ is of exponent zero and the extension is central.

The reader should consult Section 5 for our terminology on group varieties, a knowledge of STAMMBACH [3] is not required. The principal sources for this section are BEYL [4], [5]. The concepts discussed can be viewed as a group-theoretic interpretation of varietal (co)homology groups, this aspect is pursued by LEEDHAM-GREEN [1] and STAMMBACH [2].

6.1 DEFINITION. Let $\mathfrak{B}$ be a variety, Q a group in $\mathfrak{B}$, and $\tilde{R} \hookrightarrow \tilde{F} \xrightarrow{\tau} Q$ a $\mathfrak{B}$-free presentation of Q , cf. 5.2. Define

$$(6.1) \qquad K_{\mathfrak{B}}(Q) = \mathrm{Im}\{M(\tau)\colon\ M(\tilde{F}) \to M(Q)\}$$

$$(6.2) \qquad M_{\mathfrak{B}}(Q) = \mathrm{Coker}\ M(\tau) = M(Q)/K_{\mathfrak{B}}(Q) .$$

We convince ourselves that $K_{\mathfrak{B}}(Q)$ and $M_{\mathfrak{B}}(Q)$ are independent of the choice of $\mathfrak{B}$-free presentation. Given $\mathfrak{B}$-free presentations e_1 and e_2 of Q , there exists a morphism $(\alpha,\beta,1)\colon e_1 \to e_2$ of extensions because $\tilde{F}_2 \in \mathfrak{B}$ and $\tilde{F}_1$ is $\mathfrak{B}$-free:

$$\begin{array}{cccccc}
e_1 : & \tilde{R}_1 & \hookrightarrow & \tilde{F}_1 & \xrightarrow{\tau_1} & Q \\
 & \downarrow{\scriptstyle\alpha} & & \downarrow{\scriptstyle\beta} & & \| \\
e_2 : & \tilde{R}_2 & \hookrightarrow & \tilde{F}_2 & \xrightarrow{\tau_2} & Q
\end{array} .$$

The naturality assertion of Proposition I.3.5 yields a commutative diagram

$$\begin{array}{ccc} M(\tilde{F}_1) & \xrightarrow{M(\tau_1)} & M(Q) \\ {\scriptstyle M(\beta)}\downarrow & & \| \\ M(\tilde{F}_2) & \xrightarrow{M(\tau_2)} & M(Q) \end{array} ,$$

whence $\operatorname{Im} M(\tau_1) \subseteq \operatorname{Im} M(\tau_2)$. The other inclusion holds by symmetry.

In the same vein as $M(\gamma)$ was obtained in I.3, $K_{\mathfrak{B}}(\gamma)$ is defined for each homomorphism $\gamma: Q_1 \to Q_2$ in $\mathfrak{B}$ and is functorial. Since every automorphism of Q can be lifted to an endomorphism of $\tilde{F}$, $K_{\mathfrak{B}}(Q)$ is a submodule of $M(Q)$ with respect to the Aut(Q)-action.

For clarification we here assume that one free presentation $F \twoheadrightarrow Q$ has been chosen for each group Q ; the $\mathfrak{B}$-free presentation above may be related to it as in 5.2 (this is often convenient), but need not be. The coordinate isomorphisms of I.3 handle $M(Q)$ and $K_{\mathfrak{B}}(Q)$ simultaneously.

6.2 REMARKS. a) Proposition I.3.5 gives a "relative Schur-Hopf Formula"

$$M_{\mathfrak{B}}(Q) \cong \frac{\tilde{R} \cap [\tilde{F},\tilde{F}]}{[\tilde{R},\tilde{F}]} .$$

Thus this group is isomorphic with the varietal homology groups $\mathfrak{B}_1(Q,\mathbb{Z})$ of LEEDHAM-GREEN [1] and $V(Q,\mathbb{Z})$ of STAMMBACH [3; III.1].

b) In this context the group $\operatorname{Cext}_{\mathfrak{B}}(Q,A)$ of 5.8 appears as the varietal cohomology group $\tilde{V}(Q,A)$ of STAMMBACH [3; III.1]. In the notation of 6.1, $\tilde{V}(Q,A)$ is defined as the kernel of

$$\tau^* : \operatorname{Cext}(Q,A) \to \operatorname{Cext}(\tilde{F},A) .$$

We extract the proof of this remark from STAMMBACH [3; III.3], who actually treats the more general results of Knopfmacher. Let

$[e] \in \mathrm{Cext}(Q,A)$. Then $\tau^*[e] = [e\tau]$ is defined by the diagram

$$\begin{array}{rccccc} e\tau : & A & \rightarrowtail & G^{\tau} & \twoheadrightarrow & \tilde{F} \\ & \| & & \downarrow & & \downarrow \tau \\ e : & A & \rightarrowtail & G & \twoheadrightarrow & Q \end{array}$$

as in I.(1.7). First assume that $\tau^*[e] = 0$, i.e. $e\tau$ splits. Then $G^{\tau} \cong A \times F$ and its epimorphic image G are in $\mathfrak{B}$. Conversely , if e lies in $\mathfrak{B}$, then so is $G^{\tau} \subseteq G \times F$. Since $\tilde{F}$ is $\mathfrak{B}$-free, $\pi_0 : G^{\tau} \to F$ admits a splitting.

6.3 THEOREM. Let $\mathfrak{B}$ be a group variety of exponent zero, defined by the laws V , and let $Q \in \mathfrak{B}$. For any central extension

(6.3) $$e = (\varkappa,\pi) : A \rightarrowtail G \twoheadrightarrow Q ,$$

the verbal subgroup VG is the image of the composite map

$$K_{\mathfrak{B}}(Q) \hookrightarrow M(Q) \xrightarrow{\theta_*(e)} A \xrightarrow{\varkappa} G .$$

Thus $e \in \mathfrak{B}$ precisely when $\mathrm{Ker}\ \theta_*(e)$ contains $K_{\mathfrak{B}}(Q)$.

PROOF. Choose a free presentation $\rho : F \twoheadrightarrow Q$ of Q and construct the commutative diagram (6.4) from it,

(6.4) $$\begin{array}{rccccc} & VF & \hookrightarrow & F & \twoheadrightarrow & F/VF \\ & \downarrow & & \| & & \downarrow \tau \\ e' : & R & \hookrightarrow & F & \xrightarrow{\pi\rho} & Q \\ & \downarrow \sigma & & \downarrow \rho & & \| \\ e : & A & \xrightarrow{\varkappa} & G & \xrightarrow{\pi} & Q \end{array} .$$

Since $Q \in \mathfrak{B}$ and $\mathfrak{B}$ has exponent 0 , we have $VF \subseteq R \cap [F,F]$. We evaluate $M(Q)$ and $M(V/VF)$ at the free presentations exhibited by (6.4) and obtain

$$K_{\mathfrak{B}}(Q) = \mathrm{Im}\ M(\tau) = \frac{VF\cdot[R,F]}{[R,F]}$$

and then by I.3.6 and 5.1:

$$\varkappa\theta_*(e)\, K_{\mathfrak{B}}(Q) = \rho(VF\cdot[R,F]) = \rho(VF) = VG\ . \quad \square$$

6.4 COMMENTS. a) The question whether $e \in \mathfrak{B}$ or not, depends only on the subgroup $U(e)$ of $M(Q)$, cf. (3.5). We conclude that any two generalized representation groups G_1 and G_2 of Q generate the same variety of exponent zero. - For the proof, combine Theorem 6.3 for $\mathfrak{B}_1 = \mathrm{varo}(G_1)$ resp. $\mathfrak{B}_2 = \mathrm{varo}(G_2)$ with Proposition II.2.13. (This topic is resumed in 7.26.)

b) In some important cases (see 6.5 and 6.12 below) we have formulas for $K_{\mathfrak{B}}(Q)$ that do not involve free presentations.

c) Again by Theorem 6.3, $K_{\mathfrak{B}}(Q)$ essentially is the verbal subgroup of any generalized representation group of Q. The information on $K_{\mathfrak{B}}(Q)$ so obtained can then be used for arbitrary central extensions by Q.

We are going to discuss situations where $K_{\mathfrak{B}}(Q)$ can be described internally, thus eliminating the use of free presentations.

6.5 PROPOSITION. Let $\mathfrak{W}$ be a variety, defined by the laws W, let $\mathfrak{B}$ be the variety of center-by-$\mathfrak{W}$ groups and $Q \in \mathfrak{B}$. Then

$$K_{\mathfrak{B}}(Q) = \mathrm{Ker}\{M(\mathrm{nat})\colon M(Q) \to M(Q/WQ)\} = \mathrm{Im}\ \chi(e')$$

where $e'\colon WQ \hookrightarrow Q \twoheadrightarrow Q/WQ$.

Under the same assumptions $M_{\mathfrak{B}}(Q) \cong \mathrm{Im}\, M(\mathrm{nat}) = \mathrm{Ker}\ \theta_*(e')$.

PROOF. We start with a free presentation $F \twoheadrightarrow Q$, obtain a $\mathfrak{B}$-free presentation of Q as in (5.1) and construct the commutative diagram

$$
\begin{array}{ccccc}
VF & \hookrightarrow & F & \overset{\rho}{\twoheadrightarrow} & \tilde{F} \\
\downarrow & & \| & & \downarrow \tau \\
R & \hookrightarrow & F & \twoheadrightarrow & Q \\
\downarrow & & \| & & \downarrow \text{nat} \\
R\cdot WF & \hookrightarrow & F & \twoheadrightarrow & Q/WQ
\end{array} \quad .
$$

The definitions of $M(\tau)$ and $M(\text{nat})$ in terms of these free presentations give

$$\operatorname{Im} M(\tau) = \frac{(VF \cap [F,F])\cdot[R,F]}{[R,F]} \quad , \quad \operatorname{Ker} M(\text{nat}) = \frac{R \cap [R\cdot WF,F]}{[R,F]} \quad .$$

Under the present assumptions, $VF = [F,WF] \subseteq R$. Hence $(VF \cap [F,F])\cdot[R,F] = VF\cdot[R,F] \subseteq R \cap [R\cdot WF,F]$ and $R \cap [R\cdot WF,F] \subseteq$ $\subseteq [R\cdot WF,F] \subseteq [R,F]\cdot[WF,F] = VF\cdot[R,F]$. Finally, as Q is a center-by-$\mathfrak{B}$ group, e' is central; now Proposition I.3.5 yields

$$M_{\mathfrak{B}}(Q) = M(Q)/K_{\mathfrak{B}}(Q) \cong \operatorname{Ker} \theta_*(e') \ ,$$

while Theorem I.4.4 gives $\operatorname{Im} \chi(e') = \operatorname{Ker} M(\text{nat})$. □

6.6 COROLLARY (cf. EVENS [1; §3]). Let Q be nilpotent of class n and $K = \operatorname{Ker}\{M(\text{nat})\colon M(Q) \to M(Q/\Gamma_n Q)\}$. If e as in (6.3) is central, then $\Gamma_{n+1}G = \kappa\, \theta_*(e)K$. Thus G is nilpotent of class n rather $(n+1)$ precisely when $\operatorname{Ker} \theta_*(e) \supseteq K$. □

Here $\Gamma_1 Q = Q$, $\Gamma_2 Q = [Q,Q]$, $\Gamma_3 Q = [\Gamma_2 Q,Q]$.... denotes the lower central series. The assumption of the corollary is $\Gamma_n Q \neq 0$, $\Gamma_{n+1}Q = 0$.

In his study of the Dimension Conjecture, PASSI [1; p.27] introduced the notion of induced-central extensions.

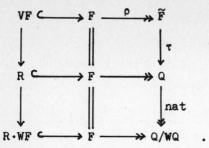

6.7 DEFINITION. Let Q be a nilpotent group of class n . A central extension $e_1\colon Q/Z \rightarrowtail M \twoheadrightarrow Q$ is called <u>induced-central</u>, if there exists a central extension e as in (6.3) and a homomorphism $f\colon A \to Q/Z$ with $\varkappa A = \Gamma_{n+1}G$ and $fe \equiv e_1$. We regard "induced-central" as a property of the congruence class $[e_1] \in \mathrm{Cext}(Q,Q/Z)$ rather then of the extension e_1. (If we wish, $\varkappa$ may be considered an inclusion map: replace A by $\varkappa A$ and f by $f\varkappa^{-1}$.)

6.8 PROPOSITION. Let Q be a nilpotent group of class $n \geq 1$ and $K = \mathrm{Ker}\{M(\mathrm{nat})\colon M(Q) \to M(Q/\Gamma_n Q)\}$. Then $e \in \mathrm{Cext}(Q,Q/Z)$ is induced-central precisely when $K + \mathrm{Ker}\ \theta_*(e) = M(Q)$.

This proposition has also been obtained by VERMANI [1] who generalized it to other kernels B instead of Q/Z. (Our proof is clearly valid for any divisible abelian group B.)

PROOF. If e as in (6.3) is any central extension then, by 6.6, $\varkappa A = \Gamma_{n+1}G$ precisely when $\theta_*(e)K = A$; this implies $K + \mathrm{Ker}\ \theta_*(e) = M(Q)$. Now assume that e_1 is induced-central and let e,f be as required. Then $e_1 \equiv fe$ implies $\theta_*(e_1) = f\,\theta_*(e)$; thus $\mathrm{Ker}\ \theta_*(e_1) \supseteq \mathrm{Ker}\ \theta_*(e)$ and $K + \mathrm{Ker}\ \theta_*(e_1) = M(Q)$. Conversely, given e_1 with $K + \mathrm{Ker}\ \theta_*(e_1) = M(Q)$. Let $A = \mathrm{Im}\ \theta_*(e_1) \subseteq Q/Z$. Decompose $\theta_*(e_1)\colon M(Q) \to Q/Z$ as $\theta_*(e_1) = f\circ g$ where $f\colon A \to Q/Z$ is the inclusion map. By the Universal Coefficient Theorem I.3.8, choose any $e \in \mathrm{Cext}(Q,A)$ with $\theta_*(e) = g$. By construction, $\mathrm{Ker}\ g = \mathrm{Ker}\ \theta_*(e_1)$ and g is epimorphic, hence $\theta_*(e)K = A$ and $\varkappa A = \Gamma_{n+1}G$. Moreover, $\theta_*(fe) = f\circ g = \theta_*(e_1)$ and $\theta_*\colon \mathrm{Cext}(Q,Q/Z) \to \mathrm{Hom}(M(Q),Q/Z)$ is an isomorphism, since Q/Z is divisible. We conclude $fe \equiv e_1$ or, that e_1 is induced-central. □

6.9 COROLLARY. Let Q be nilpotent of class $n \geq 1$. Then every $e \in \mathrm{Cext}(Q,Q/Z)$ is induced-central if and only if,

$M(\mathrm{nat})\colon M(Q) \to M(Q/\Gamma_n Q)$ is the zero-map.

PROOF. If $K \neq M(Q)$, there is a non-trivial homomorphism $M(Q)/K \to \mathbb{Q}/\mathbb{Z}$, because $\mathbb{Q}/\mathbb{Z}$ is injective (divisible). We obtain a non-zero homomorphism $f\colon M(Q) \to \mathbb{Q}/\mathbb{Z}$ with $f|_K = 0$. By Proposition 6.8, none of $e \in \mathrm{Cext}(Q,\mathbb{Q}/\mathbb{Z})$ with $\theta_*(e) = f$ is induced-central. The converse is immediate. □

6.10 COROLLARY (PASSI [1; Thm. 4.2]). Let Q be nilpotent of class n such that $\Gamma_n Q$ has finite exponent m. Then $m[e] = 0$ for all induced-central extensions $[e] \in \mathrm{Cext}(Q,\mathbb{Q}/\mathbb{Z})$.

PROOF. Let e be induced-central and K as in Proposition 6.8. Applying Proposition 6.5 for $\mathfrak{B} = \mathfrak{N}_{n-1}$, we obtain

$$K = \mathrm{Im}\{\chi(e')\colon Q_{ab} \otimes \Gamma_n Q \to M(Q)\},$$

thus K has exponent dividing m. Now $K + \mathrm{Ker}\,\theta_*(e) = M(Q)$ and $mK = 0$ imply $\theta_*(me) = m\theta_*(e) = 0$. Since θ_* is an isomorphism as in the proof of 6.8, $m[e] = [me] = 0$. □

It has been an apparently open problem whether the induced-central extensions form a subgroup of $\mathrm{Cext}(Q,\mathbb{Q}/\mathbb{Z})$. Although Corollary 6.10 suggests the opposite, the answer is negative.

6.11 EXAMPLE. In general, the subset of (congruence classes of) induced-central extensions in $\mathrm{Cext}(Q,\mathbb{Q}/\mathbb{Z})$ is not a subgroup. Let $n = 2$, D be the dihedral group of order 8, $Q = D \times \mathbb{Z}/2$, and $\pi\colon Q \twoheadrightarrow Q_{ab} \cong (\mathbb{Z}/2)^3$ denote the abelianization. We use $M(D) \cong \mathbb{Z}/2$ from Example II.3.8(iv). Then

$$M(Q) \cong M(Q_{ab}) \cong (\mathbb{Z}/3)^3$$

by the Schur-Künneth Formula II.(4.7) and Theorem I.4.7. Counting orders in the exact sequence I.(3.3') belonging to π, we obtain

$K = \text{Ker } M(\pi) \cong \mathbb{Z}/2$. One easily finds homomorphisms $f_1, f_2 \colon M(Q) \to$ $\to \mathbb{Z}/2 \hookrightarrow \mathbb{Q}/\mathbb{Z}$ with $f_1 + f_2 \neq 0$ and $\text{Ker}(f_1+f_2) \supseteq K$ and $K \cap \text{Ker } f_1 = K \cap \text{Ker } f_2 = 0$. Choose $e_i \in \text{Cext}(Q, \mathbb{Q}/\mathbb{Z})$ with $\theta_*(e_i) = f_i$ for $i := 1,2$. Then e_1 and e_2 are induced-central by Proposition 6.8, but $e_1 + e_2$ is not.

6.12 PROPOSITION. Let $\mathfrak{W}$ be a variety, defined by the laws W , let $\mathfrak{B}$ be the variety of abelian-by-$\mathfrak{W}$ groups, and let $Q \in \mathfrak{B}$. Then $K_{\mathfrak{B}}(Q) = \text{Im}\{\text{res}\colon M(WQ) \to M(Q)\}$ with $\text{res} = M(\text{incl})$.

PROOF. Again, we construct a commutative diagram

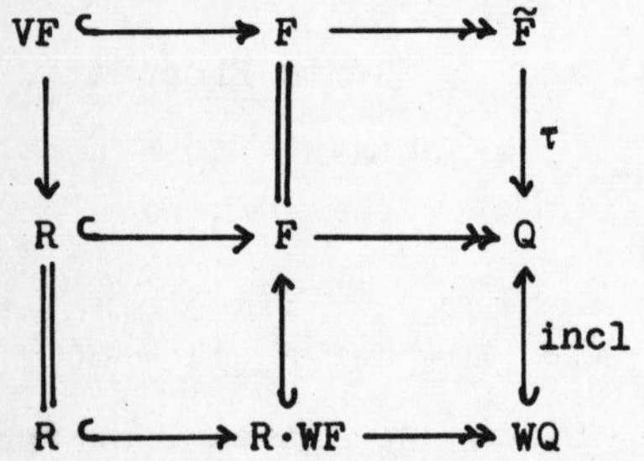

with F free, hence $R \cdot WF$ free and $\tilde{F}$ $\mathfrak{B}$-free. By I.3.2:

$$\text{Im } M(\tau) = \frac{[WF,WF]\cdot[R,F]}{[R,F]} \, , \quad \text{Im}(\text{res}) = \frac{(R \cap [R\cdot WF, R\cdot WF])\cdot[R,F]}{[R,F]} \, .$$

Now $VF = [WF,WF] \subseteq R$, thus $[WF,WF]\cdot[R,F] \subseteq (R \cap [R\cdot WF, R\cdot WF])\cdot[R,F]$ and $(R \cap [R\cdot WF, R\cdot WF])\cdot[R,F] \subseteq [R\cdot WF, R\cdot WF]\cdot[R,F] \subseteq [WF,WF]\cdot[R,F]$. □

6.13 COROLLARY. If Q is solvable of length $l \geq 1$ and e as in (6.3) is a central extension, then G is solvable of length l rather then $(l+1)$ precisely when

$$\text{Ker } \theta_*(e) \supseteq \text{Im}\{M(D_{l-1}Q) \to M(Q)\} \, .$$

Here $D_{l-1}Q$ denotes the last non-trivial term of the derived series of Q . □

For finite groups, a cohomological formulation of this corollary is due to YAMAZAKI [1; §3.2].

6.14 EXAMPLE. Let G be a group such that $[G,G] \neq 0$ is abelian with trivial multiplicator, e.g. cyclic. Then every central extension by G is solvable of length two. Indeed, take $\mathfrak{W} = \mathfrak{S}_1$ and $\mathfrak{V} = \mathfrak{S}_2$, then $K_{\mathfrak{V}}(G) = 0$ by Corollary 6.13.

6.15 DEFINITION. Let $\mathfrak{V}$ be a variety and $Q \in \mathfrak{V}$. Then Q is called <u>absolutely-$\mathfrak{V}$</u>, if every central extension $A \rightarrowtail G \twoheadrightarrow Q$ by Q again lies in $\mathfrak{V}$.

6.16 THEOREM. Let $\mathfrak{V}$ be a variety of exponent 0 and $Q \in \mathfrak{V}$. Then the following are equivalent:

(i) Q is absolutely-$\mathfrak{V}$;

(ii) all generalized representation groups of Q lie in $\mathfrak{V}$;

(iii) some generalized representation group of Q lies in $\mathfrak{V}$;

(iv) $K_{\mathfrak{V}}(Q) = 0$.

By the very definition, absolutely-$\mathfrak{V}$ groups can exist only if $\mathfrak{V}$ has exponent 0 . However, an arbitrary variety $\mathfrak{V}$ determines the variety $\mathfrak{W} = \mathrm{varo}(\mathfrak{V})$ of exponent 0 , cf. 5.7. Then one has $K_{\mathfrak{V}}(Q) = K_{\mathfrak{W}}(Q)$ for $Q \in \mathfrak{V}$; the influence of $K_{\mathfrak{V}}(Q)$ is described by the theorem for $\mathfrak{W}$.

PROOF. Let e as in (6.3) be a central extension. Then $VG \cong \theta_*(e)K_{\mathfrak{V}}(Q)$ by Theorem 6.3. If e is a generalized representation group, then $\theta_*(e)$ is a monomorphism by Proposition II.2.13, hence $G \in \mathfrak{V}$ implies $K_{\mathfrak{V}}(Q)=0$. The remaining assertions of the theorem follow easily. Finally let $Q \in \mathfrak{V}$ and $\mathfrak{W} = \mathrm{varo}(\mathfrak{V})$. Then the formula $K_{\mathfrak{V}}(Q) = K_{\mathfrak{W}}(Q)$ follows from $WF = VF \cap [F,F]$ for free groups F , cf. 5.7. □

6.17 EXAMPLES. a) Let $\mathfrak{B}$ be a variety of exponent 0 and $Q \in \mathfrak{B}$ with $M(Q) = 0$; then Q is absolutely-$\mathfrak{B}$ by Theorem 6.3. The groups of Example 6.14 are absolutely-$\mathfrak{S}_2$.

b) By various reasons (use Corollary I.3.9 or Example 4.8 or Theorem 6.16) an abelian group Q is absolutely-abelian precisely when $M(Q) = 0$. VARADARAJAN [1] classified the abelian groups with trivial multiplicator in a topological context, while MOSKALENKO [1] classified the absolutely-abelian groups as such. Finally BEYL [3], among other things, obtained the absolutely-abelian groups via the multiplicator. The abelian groups Q in question are characterized by the total of the following properties (Q_p denotes the Sylow p-subgroup of Tor Q) :

(i) the rational rank of Q does not exceed one ;

(ii) the reduced part of Q_p is 0 or cyclic ;

(iii) for each prime p , Q/TorQ is p-divisible unless $Q_p = 0$.

6.18 EXAMPLES. a) A group Q is absolutely-$\mathfrak{N}_2$ precisely when $[Q,Q] \subseteq Z^*(Q)$. Indeed, by 3.7(b) every representation group G of Q has $G/Z(G) \cong Q/Z^*(Q)$ abelian; the assertion follows by Theorem 6.16.

b) A nilpotent torsion group Q is absolutely-$\mathfrak{N}_2$ if, and only if, so are all its Sylow p-subgroups. For Q and [Q,Q] and Z*(Q) by Theorem 4.2 are the restricted direct products of their Sylow subgroups, see also 4.3.

c) We describe the metacyclic groups G that are absolutely-$\mathfrak{N}_2$. Let $G = G(m,n,r,\lambda)$ as in (2.2), the parameters being subject to (2.3) and (2.5). By Proposition 4.18 Z*(G) is the cyclic group generated by b^k , where k is a certain integer dividing n .

Since $[G,G]$ is generated by $a^{(m,r-1)}$ and $b^{\nu} \notin \langle a \rangle$ for $0 < \nu < n$ and $\lambda | (m,r-1)$, the condition $[G,G] \subseteq Z^*(G)$ amounts to

$$\frac{m}{(m,r-1)} \mid \frac{(m,r-1)}{\lambda} \quad \text{resp.} \quad \lambda \mid \frac{1}{m}(m,r-1)^2 .$$

7. Schur-Baer Multiplicators and Isologism.

Here we demonstrate that some methods of Sections I.3 and III.2 easily adapt from isoclinism to isologism. The isologism relation puts groups resp. extensions into the same class roughly if they equally deviate from lying in a given variety $\mathfrak{V}$. For the moment, a reasonable level of understanding is achieved by assuming that $\mathfrak{V}$ is defined by a single law $v \in F_\infty$. For example, the law $v = [x_1,x_2]$ defines $\mathfrak{A}$, while the corresponding concept of $\mathfrak{A}$-isologism agrees with isoclinism. In general, the center $Z(G)$ is replaced by the marginal subgroup $v^*(G)$ of those elements of G which are "not noticed" by v , $M(Q)$ is replaced by the Schur-Baer multiplicator $\mathfrak{V}M(Q)$. This section owes a great deal to P. HALL [1], [2] and LEEDHAM-GREEN/McKAY [1]. We hope that the present treatment may serve as an introduction to the latter paper which eventually focusses on certain varietal cohomology groups (different from those considered in Section 6). Our elementary approach is limited insofar as $\mathfrak{V}\theta_*$, the analogue of the projection map in the Universal Coefficient Theorem, is not surjective in general, cf. LEEDHAM-GREEN/McKAY [1; II §3].

7.1 DEFINITION (cf. P. Hall [2]). Let $V \subseteq F_\infty$ be a set of laws and $v = v(x_1,\ldots,x_n)$ be a typical element of V , with n depending on v . For every group G , consider the subset of G given by the elements x satisfying

$$v(g_1,\ldots,g_i\cdot x,\ldots,g_n) = v(g_1,\ldots,g_i,\ldots,g_n)$$

for all $g_j \in G$, all places i . One easily verifies that this subset $v^*(G)$ is a characteristic subgroup of G , cf. (5.3). The V-marginal subgroup of G is defined as

$$V^*G = \bigcap_{v \in V} v^*(G) \quad .$$

By the very definition of V^*G, every law as above determines a functor $\tilde{v}$ from groups to set functions which assigns to G the function

$$\tilde{v}_G : (G/V^*G)^n := \prod^n (G/V^*G) \longrightarrow VG$$

given by v.

7.2 DEFINITION. A group extension

$$(7.1) \qquad e : N \overset{\varkappa}{\rightarrowtail} G \overset{\pi}{\twoheadrightarrow} Q$$

is called a V-<u>marginal</u> extension, if $\varkappa N \subseteq V^*G$. Such an extension and every law v in V give rise to a set function $\tilde{v}_e : Q^n \to VG$, natural with respect to morphisms of V-marginal extensions. For an arbitrary extension e, consider the subgroup T of VG generated by all

$$v(g_1,\ldots,g_i\cdot\varkappa(x),\ldots,g_n)\cdot v(g_1,\ldots,g_i,\ldots,g_n)^{-1}$$

with $v \in V$, $g_j \in G$, $x \in N$, all places i. Then $T \subseteq \varkappa N$ since π maps every generator of T onto 1. We denote T by $[NV^*G]_\varkappa$ or simply by $[NV^*G]$. Hence e is V-marginal precisely when $[NV^*G]_\varkappa = 0$.

7.3 REMARK. It is quite practical to know that $V^*G = W^*G$ where $W = VF_\infty$ is the associated closed set of laws. Hence we may unambiguously speak of the <u>$\mathfrak{V}$-marginal subgroup</u> of G, where $\mathfrak{V}$ is the variety defined by the laws V.

PROOF. Clearly $W^*(G) \subseteq V^*(G)$. It is easy to show $U^*(G) = V^*(G)$ where U is the subgroup rather than fully invariant subgroup of F_∞ generated by V. Let $v = v(x_1,\ldots,x_n) \in V$ and let f be any endomorphism of F_∞. We consider the law

$$w = f(v) = v(f(x_1),\dots,f(x_n)) ,$$

cf. (5.3). Let m be large enough such that $\{x_1,\dots,x_m\}$ contains all variables that appear in any of the words $y_1 = f(x_1),\dots,y_n = f(x_n)$; we treat the y_j and w as words in the variables $x_1,\dots,x_m$. Recall $V^*(G) \trianglelefteq G$ from 7.1. Appeal to the natural projection $G \twoheadrightarrow G/V^*(G)$ shows the following: If $g_1,\dots,g_m,g_1',\dots,g_m' \in G$ and $g_i' \equiv g_i \mod V^*(G)$ for $i = 1,\dots,m$, then

$$y_j(g_1',\dots,g_m') \equiv y_j(g_1,\dots,g_m) \mod V^*(G)$$

for $j = 1,\dots,n$. The defining property of $V^*(G)$ now gives

$$\begin{aligned} w(g_1',\dots,g_m') &= v(y_1(g_1',\dots,g_m'),\dots) \\ &= v(y_1(g_1,\dots,g_m),\dots) = w(g_1,\dots,g_m) . \end{aligned}$$

This means $V^*(G) \subseteq w^*(G)$, what was to be shown. □

In the same vein, we may speak of a $\mathfrak{V}$-marginal rather than V-marginal extension. Also

(7.2) $\mathfrak{V}G = 0$ iff $G \in \mathfrak{V}$ iff $\mathfrak{V}^*G = G$.

For the remainder of this section, let us assume that the laws V are given and $\mathfrak{V}$ is the group variety defined by V or, equivalently, a group variety $\mathfrak{V}$ is given and V is a generating set of laws. The following is rather immediate from the definition:

7.4 LEMMA. With the notation of 7.2, $[NV^*G]$ is a normal subgroup of G yielding a commutative diagram

(7.3)
$$\begin{array}{rccccc} e : & N & \overset{\varkappa}{\rightarrowtail} & G & \overset{\pi}{\twoheadrightarrow} & Q \\ & \downarrow \text{nat} & & \downarrow \text{nat} & & \| \\ m(e) : & \dfrac{N}{\varkappa^{-1}[NV^*G]} & \overset{\varkappa'}{\rightarrowtail} & \dfrac{G}{[NV^*G]} & \overset{\pi'}{\twoheadrightarrow} & Q \end{array} ,$$

where the vertical maps are natural projections and the horizontal maps are induced by $\varkappa$ and π . The extension $m(e)$ is $\mathfrak{V}$-marginal. Any morphism $e \to e_1$ in a $\mathfrak{V}$-marginal extension e_1 factors uniquely over $e \to m(e)$. Hence any morphism $(\alpha,\beta,\gamma)\colon e_1 \to e_2$ induces a morphism $(\alpha_m,\cdot,\gamma)\colon m(e_1) \to m(e_2)$, functorially. □

Consequently $[NV^*G]$ is the least normal subgroup T of G such that e/T is $\mathfrak{V}$-marginal; in view of 7.3 the notation $[N\mathfrak{V}^*G]$ is justified.

7.5 PROPOSITION (LEEDHAM-GREEN/McKAY [1; Prop. I.1.3]). Let $\mathfrak{W}$ be any variety and $\mathfrak{V}$ consist of the center-by-$\mathfrak{W}$ groups. Then $[V^*G,WG] = 0$. In particular,

$$(7.4) \qquad [W^*G,WG] = 0 .$$

PROOF. Let $W = \mathfrak{W}F_\infty$ and $w=w(x_1,\dots,x_n) \in W$, then

$$v(x_1,\dots,x_{n+1}) = [x_{n+1},w(x_1,\dots,x_n)]$$

is a law for $\mathfrak{V}$ by 5.5. Now let $x \in V^*(G)$ and $g_1,\dots,g_n \in G$. Then

$$v(g_1,\dots,g_n,1\cdot x) = v(g_1,\dots,g_n,1) = 1 .$$

This means that x commutes with all subgroup generators, hence with all elements, of WG . Formula (7.4) holds since $\mathfrak{W} \subseteq \mathfrak{V}$ implies $W^*G \subseteq V^*G$. □

7.6 EXAMPLE (P. HALL [2]). Let $\mathfrak{V} = \mathfrak{N}_c$ with $c \geq 1$. Then $\mathfrak{V}G = \Gamma_{c+1}G$ and $\mathfrak{V}^*(G) = Z_c(G)$. Here $Z_o(G) = 0$, $Z_1(G) = Z(G)$, $Z_2(G),\dots$ denotes the upper central series of G .

PROOF. The formula $VG = \Gamma_{c+1}G$ follows from 5.5 and 5.4. We first prove $V^*(G) \subseteq Z_c(G)$. Let $x \in V^*(G)$, while $g_1,g_2,\dots$ are arbitrary elements of G . Then

$$[g_1,\ldots[g_c,1\cdot x]\ldots] = [g_1\ldots[g_c,1]\ldots] = 1$$

implies $[g_2,\ldots[g_c,x]\ldots] \in Z_1(G)$, then $[g_3,\ldots[g_c,x]\ldots] \in Z_2(G)$, and finally $x \in Z_c(G)$. We are left to prove $Z_c(G) \subseteq V^*(G)$. The case $c = 1$ is clear, we proceed by induction for $c \geq 2$. Note that $\Gamma_{c+1}F_\infty$, as a fully invariant subgroup of F_∞ , is generated by the single law

(7.5) $\quad v_c(x_1,\ldots,x_{c+1}) = [x_1,\ldots[x_c,x_{c+1}]\ldots]$,

cf. HUPPERT [1; p.257]. Let $x \in Z_c(G)$, thus $xZ(G) \in Z_{c-1}(G/Z(G))$. Then

$$v_{c-1}(g_2,\ldots,g_i\cdot x,\ldots,g_{c+1}) \equiv v_{c-1}(g_2,\ldots,g_i,\ldots,g_{c+1}) \mod Z(G)$$

by the induction step, hence

$$v_c(g_1,\ldots,g_i\cdot x,\ldots,g_{c+1}) = v_c(g_1,\ldots,g_i,\ldots,g_{c+1})$$

for $2 \leq i \leq c+1$. The missing case $i = 1$ is by Lemma I.4.1(a) a consequence of $v_c(x,g_2,\ldots,g_{c+1}) = 1$. From I.4.1(e) with $y := g_2^{-1}$ and $z := [g_3,\ldots,g_{c+1}]$ resp. $z := g_3$ for $c = 2$, we obtain

$$v_c(x,g_2,\ldots,g_{c+1}) = {}^{y^{-1}x}[[x^{-1},y],z]\cdot{}^{y^{-1}z}[[z^{-1},x],y] .$$

The first factor on the right side vanishes by induction since $[x^{-1},y] \in Z_{c-1}(G)$. The second factor vanishes since the induction hypothesis for $G/Z(G)$ implies $[x,z] \equiv 1 \mod Z(G)$. □

7.7 EXAMPLE. Again let $\mathfrak{B} = \mathfrak{N}_c$ for $c \geq 1$. The fact that $v_c(x_1,\ldots,x_{c+1})$ as in (7.5) generates $\mathfrak{B}F_\infty$, also implies

$$[N\mathfrak{B}^*G] = [G,\ldots[G,N]\ldots]$$

with c entries G , for $N \lhd G$. It is clear from the definition that $[N\mathfrak{B}^*G]$ contains $[G,\ldots[G,N]\ldots]$. For the proof of equality one may proceed by induction on c . The tool is Lemma I.4.1(a),(b) and a typical argument is as follows: $x \equiv y \mod [G\ldots,N]$ implies

$[g,x] \equiv [g,y] \mod [G,[G\ldots,N]]$ by I.4.1(b).

We are now going to define the Schur-Baer multiplicator of any group Q, possibly not in $\mathfrak{B}$, and to establish an exact sequence in analogy to Proposition I.3.5.

7.8 PROPOSITION. There is a functor s' from the category of extensions to that of central extensions which assigns to e as in (7.1) the extension

$$s'(e): \frac{\varkappa N \cap VG}{[NV^*G]} \hookrightarrow \frac{VG}{[NV^*G]} \xrightarrow{\pi_*} VQ ,$$

and to morphisms of extensions the obvious induced maps. If $(1,\beta,1)\colon e_1 \to e_2$ is a congruence, then so is $s'(1,\beta,1)$. If $(\alpha,\beta,\gamma),(\alpha',\beta',\gamma)\colon e_1 \to e_2$ are given, then $s'(\alpha',\beta',\gamma) = s'(\alpha,\beta,\gamma)$. In this sense, $s'(\alpha,\beta,\gamma)$ depends on γ only.

PROOF. The functor s' is a composite of functors: first m of 7.4, then the restriction to the V-verbal subgroups. Since $m(e)$ is V-marginal, $s'(e)$ is central by (7.4). If $(1,\beta,1)$ is a congruence, then one shows $\varkappa_1^{-1}[NV^*G_1] = \varkappa_2^{-1}[NV^*G_2]$, thus $s'(1,\beta,1)$ is a congruence. Given $(\alpha,\beta,\gamma),(\alpha',\beta',\gamma)\colon e_1 \to e_2$, then $\beta'(g) = \varkappa_2(n_g)\cdot\beta(g)$ for all $g \in G_1$, with $n_g \in N_2$. A typical generator of $V(G_1)/[N_1V^*G_1]$ has the form $x = v(g_1,\ldots,g_n)\cdot[N_1V^*G_1]$, and $\beta'(v(g_1,\ldots,g_n))\cdot\beta(v(g_1,\ldots,g_n))^{-1} = v(\beta'(g_1),\ldots,\beta'(g_n))\cdot$ $\cdot v(\beta(g_1),\ldots,\beta(g_n))^{-1} \in [N_2V^*G_2]$. Hence β' and β induce the same map. □

7.9 DEFINITION. If $e\colon R \hookrightarrow F \twoheadrightarrow Q$ is a free presentation of the group Q, then define the value of the <u>Schur-Baer multiplicator</u> at the coordinate system e as the abelian group $\mathfrak{B}M(Q)_e =$ $= (R \cap VF)/[RV^*F]$. Given any homomorphism $\gamma\colon Q \to Q'$ of groups and free presentations e and e' of Q and Q', respectively, define $\mathfrak{B}M(\gamma)_{e|e'} = s'(\alpha,\beta,\gamma)$ for any choice of lifting (α,β,γ) .

By Proposition 7.8, $\mathfrak{B}M(\gamma)_{e|e'}$ is well-defined and the analogue of I.(3.1) holds for $\mathfrak{B}M(-)$; there are unique coordinate transformations. Certainly, $\mathfrak{A}M(Q)_e = M(Q)_e$, as defined in I.3.2. Incidentally, the same argument gives the invariance of $\mathfrak{B}P(Q) :=$ $:= VF/[RV^*F]$ up to coordinate isomorphisms. Both $\mathfrak{B}M(Q)$ and $\mathfrak{B}P(Q)$ are special Baer-invariants, see LEEDHAM-GREEN/McKAY [1; I §1] for more details.

7.10 PROPOSITION. Every extension e as in (7.1) determines an exact sequence

$$(7.6)\quad \mathfrak{B}M(G) \xrightarrow{\mathfrak{B}M(\pi)} \mathfrak{B}M(Q) \xrightarrow{\mathfrak{B}\theta_*(e)} \frac{N}{\varkappa^{-1}[NV^*G]} \xrightarrow{\varkappa_*} \frac{G}{VG} \xrightarrow{\pi_*} \frac{Q}{VQ} \longrightarrow 0 ,$$

which is natural with respect to morphisms of extensions. (Here $\varkappa_*$ and π_* are induced by $\varkappa$ and π). If $e \equiv e'$, then $\mathfrak{B}\theta_*(e) =$ $= \mathfrak{B}\theta_*(e')$. If $G = F$ is free and $\varkappa$ an inclusion, then $\mathfrak{B}\theta_*(e)$ is just the inclusion

$$\mathfrak{B}M(Q)_e = (N \cap VF)/[NV^*F] \hookrightarrow N/[NV^*F] .$$

In the special case that e is $\mathfrak{B}$-marginal, i.e. $[NV^*G] = 0$, FRÖHLICH [1; Thm. 3.2] obtained the sequence (7.6) in the context of algebra varieties. (Actually, our proof can be easily adapted to varieties of algebras.)

SKETCH OF PROOF. The proof is parallel to that of Proposition I.3.5. Starting with a free presentation I.(3.4) of e , we consider the sequence

$$\frac{S \cap V(F)}{[SV^*F]} \longrightarrow \frac{R \cap V(F)}{[RV^*F]} \xrightarrow{\theta'} \frac{R}{[RV^*F]\cdot S} \longrightarrow \frac{F}{S\cdot V(F)} \longrightarrow \frac{F}{R\cdot V(F)} .$$

With $\mathfrak{B}M(Q)_{e'} = (R \cap V(F))/[RV^*F]$ and $\sigma_*: R/([RV^*F]\cdot S) \cong$ $\cong N/\varkappa^{-1}[NV^*G]$, we define $\mathfrak{B}\theta_*(e) = \sigma_*\circ\theta' : \mathfrak{B}M(Q)_{e'} \to N/\varkappa^{-1}[NV^*G]$.

One shows that different choices of I.(3.4) give compatible descriptions of $\mathfrak{B}\theta_*(e)$. The exactness of the above sequence follows by the modular law from $S \subseteq R$ and $[RV^*F] \subseteq R \cap V(F)$. □

7.11 REMARKS. (a) The naturality assertion of Proposition 7.10 yields the following description of $\mathfrak{B}\theta_*(e)$. Let e be the free presentation of Q at which $\mathfrak{B}M(Q)$ shall be evaluated, obtain $(\alpha,\beta,1)\colon \bar{e} \to e$ like I.(3.5); then $\mathfrak{B}\theta_*(e)$ is the composite

$$\mathfrak{B}M(Q)_{\bar{e}} = \frac{R \cap V(F)}{[RV^*F]} \xrightarrow{\mathfrak{B}\theta_*(\bar{e})} \frac{R}{[RV^*F]} \xrightarrow{\alpha_m} \frac{N}{\varkappa^{-1}[NV^*G]} .$$

Naturality also implies $\mathfrak{B}\theta_*(e) = \mathfrak{B}\theta_*(m(e))$ and $\mathfrak{B}\theta_*(e) = \mathfrak{B}\theta_*(e')$ for $e \equiv e'$.

(b) The group Q acts on itself by inner automorphisms $i_q = \{x \longmapsto qxq^{-1}\}\colon Q \to Q$. Hence by the functoriality of $\mathfrak{B}M(-)$, Q acts on $\mathfrak{B}M(Q)$ by $q\cdot y = M(i_q)y$ for $q \in Q$ and $y \in \mathfrak{B}M(Q)$. If e is an extension by Q , then every i_q can be lifted to a map $(\alpha,\beta,i_q)\colon e \to e$ with β an automorphism; thus Ker $\mathfrak{B}\theta_*(e)$ is a Q-submodule of $\mathfrak{B}M(Q)$.

(c) For every Q-submodule K of $\mathfrak{B}M(Q)$, there is a $\mathfrak{B}$-marginal extension e by Q with Ker $\mathfrak{B}\theta_*(e) = K$. To this end, start with a free presentation $\bar{e}_0\colon R \hookrightarrow F \twoheadrightarrow Q$, form $m(\bar{e})$, and regard K as a subgroup of $F/[RV^*F]$ via $\mathfrak{B}M(Q)_{\bar{e}} = (R \cap VF)/[RV^*F]$. Since K is a Q-submodule of $\mathfrak{B}M(Q)$, K is normal in $F/[RV^*F]$. Then $e_1 = m(\bar{e})/K$ is a $\mathfrak{B}$-marginal extension with Ker $\mathfrak{B}\theta_*(e_1) = K$.

7.12 REMARK. Now let e as in (7.1) be $\mathfrak{B}$-marginal. Then $[NV^*G] = 0$, and we regard N rather than $N/0$ as the range of $\mathfrak{B}\theta_*(e)$. We obtain from (7.6) and 5.1 the exact sequence

(7.7) $\qquad \text{Ker } \mathfrak{B}\theta_*(e) \hookrightarrow \mathfrak{B}M(Q) \xrightarrow{\theta} V(G) \longrightarrow V(Q) \longrightarrow 0$

of groups, where θ is the restriction of $\varkappa\circ\mathfrak{B}\theta_*(e)\colon \mathfrak{B}M(Q) \to N \to G$.

This sequence is natural with respect to maps of $\mathfrak{V}$-marginal extensions. Consequently, if $Q \in \mathfrak{V}$, then $G \in \mathfrak{V}$ precisely when $\mathfrak{V}\theta_*(e) = 0$.

7.13 DEFINITION. Let $e_i\colon N_i \rightarrowtail G_i \xrightarrow{\pi_i} Q_i$ for $i = 1,2$ be $\mathfrak{V}$-marginal group extensions. Then e_1 and e_2 are called $\mathfrak{V}$-<u>isologic</u>, if there are isomorphisms $\eta\colon Q_1 \to Q_2$ and $\xi\colon V(G_1) \to V(G_2)$ such that the following diagram is commutative for all $v \in V$,

$$(7.8) \qquad \begin{array}{ccc} Q_1^n & \xrightarrow{\tilde{v}_1} & V(G_1) \\ \downarrow{\scriptstyle \eta^n} & & \downarrow{\scriptstyle \xi} \\ Q_2^n & \xrightarrow{\tilde{v}_2} & V(G_2) \end{array},$$

where $\tilde{v}_i$ denotes the word function of e_i for $i = 1,2$ (see 7.2). The pair (ξ,η) is called a $\mathfrak{V}$-<u>isologism</u>. If $Q_1 = Q_2$ and there is an isologism $(\xi,1)$, this is called a <u>special</u> $\mathfrak{V}$-<u>isologism</u>.

7.14 REMARKS. (a) Every group G determines a $\mathfrak{V}$-marginal extension $e_G\colon V^*(G) \hookrightarrow G \twoheadrightarrow G/V^*(G)$. Then G and H are $\mathfrak{V}$-isologic in the sense of P. Hall [2] precisely when e_G and e_H are $\mathfrak{V}$-isologic in the sense of Def. 7.13.

(b) If e_1 and e_2 are $\mathfrak{V}$-isologic as $\mathfrak{V}$-marginal extensions, then the middle groups G_1 and G_2 are $\mathfrak{V}$-isologic as groups. Moreover, if $\varkappa_1 N_1 = V^*(G_1)$, then $\varkappa_2 N_2 = V^*(G_2)$. Indeed, if e as in (7.1) is $\mathfrak{V}$-marginal and $q \in Q$, then $q \in \pi V^*(G)$ precisely when $\tilde{v}(q_1,\dots,q_i\cdot q,\dots,q_n) = \tilde{v}(q_1,\dots,q_i,\dots,q_n)$ for all laws $v \in V$, all $q_1,\dots,q_n \in Q$, and all places i. Thus the word functions $\tilde{v}$ detect $\pi V^*(G)$ and determine $G/V^*(G) \cong Q/\pi V^*(G)$. If (ξ,η) is a $\mathfrak{V}$-isologism between e_1 and e_2, then $\eta\pi_1 V^*(G_1) = \pi_2 V^*(G_2)$, whence the assertions follow.

7.15 LEMMA. Let $f\colon G \to H$ be an epimorphism of groups with $(\operatorname{Ker} f) \cap VG = 0$. Then $f^{-1}(V^*H) = V^*G$ and $(\xi = f|_{VG,VH} : \eta = f_*\colon G/V^*G \to Q/V^*H)$ is a $\mathfrak{V}$-isologism.

PROOF. Certainly ξ is an isomorphism and $f(V^*G) \subseteq V^*H$. Now let $y \in V^*H$ and $x \in G$ with $f(x) = y$, let v be any law of V. Then $v(g_1,\dots,g_ix,\dots,g_n)\cdot v(g_1,\dots,g_i,\dots,g_n)^{-1} \in (\operatorname{Ker} f) \cap VG = 0$ for all $g_1,\dots,g_n \in G$ and all places i, hence $x \in V^*G$. Thus $f^{-1}(V^*H) = V^*G$ and $\eta = f_*$ is an isomorphism. The commutativity of (7.8), for the extensions e_G and e_H, is obvious. □

7.16 DEFINITION. An epimorphism $f\colon G \to H$ with $(\operatorname{Ker} f) \cap VG = 0$ is called a $\mathfrak{V}$-isologic epimorphism of groups. Let

$$e_i : N_i \rightarrowtail G_i \xrightarrow{\pi_i} Q_i$$

be $\mathfrak{V}$-marginal extensions for $i = 1,2$. If there exists an epimorphism $\beta\colon G_1 \to G_2$ with $(\operatorname{Ker} \beta) \cap VG = 0$ and $\beta^{-1}(\operatorname{Ker} \pi_2) = \operatorname{Ker} \pi_1$, then the uniquely induced morphism $(\alpha,\beta,\gamma)\colon e_1 \to e_2$ is called an $\mathfrak{V}$-isologic epimorphism of extensions, since $(\xi = \beta|_{VG_1,VG_2}; \eta = \gamma)$ is a $\mathfrak{V}$-isologism of extensions; if moreover $Q_1 = Q_2$ and $\gamma = 1$, we call (α,β,γ) a special $\mathfrak{V}$-isologic epimorphism.

7.17 DEFINITION. A group extension e as in (7.1) is of $\mathfrak{V}$-type if $\varkappa(N) \cap V(G) = 0$. If e is of $\mathfrak{V}$-type, then $N \in \mathfrak{V}$. If $e \equiv e'$ and e is of $\mathfrak{V}$-type, then e' is also of $\mathfrak{V}$-type. If $N \in \mathfrak{V}$, we denote by $\operatorname{EXT}_{\mathfrak{V}}(Q,N) \subseteq \operatorname{EXT}(Q,N)$ the congruence classes of $\mathfrak{V}$-type. For $A \in \mathfrak{V}\cap\mathfrak{A}$, let $\operatorname{Cext}_{\mathfrak{V}}(Q,A) = \operatorname{EXT}_{\mathfrak{V}}(Q,A) \cap \operatorname{Cext}(Q,A)$.

If $Q \in \mathfrak{V}$, then e as in (7.1) is of $\mathfrak{V}$-type precisely when $G \in \mathfrak{V}$. Thus the present definition of $\operatorname{Cext}_{\mathfrak{V}}(Q,A)$ agrees with that in 5.8 for $Q \in \mathfrak{V}$. An extension of $\mathfrak{V}$-type is essentially a $\mathfrak{V}$-isologic epimorphism; it is $\mathfrak{V}$-marginal by Lemma 7.15. The totality of $\mathfrak{V}$-isologic epimorphisms onto a given group Q is parametrized by the epimorphisms in $\mathfrak{V}$ with range $Q/\mathfrak{V}Q$.

7.18 PROPOSITION. Let $\mathfrak{V}$ be a group variety, $N \in \mathfrak{V}$, and Q an arbitrary group. Then $nat: Q \twoheadrightarrow Q/VQ$ induces a bijection

$$nat^* : EXT_{\mathfrak{V}}(Q/VQ,N) \to EXT_{\mathfrak{V}}(Q,N) ;$$

cf. I.1.4. If $N \in \mathfrak{V} \cap \mathfrak{A}$, then $Cext_{\mathfrak{V}}(Q,N)$ is a subgroup of $Cext(Q,N)$ and nat induces an isomorphism

$$nat^* : Cext_{\mathfrak{V}}(Q/VQ,N) \to Cext_{\mathfrak{V}}(Q,N) .$$

PROOF, cf. LUE [1; Prof. 1.4]. For $[e] \in EXT_{\mathfrak{V}}(Q,N)$ construct

$$\begin{array}{rccccc}
 & 0 & \rightarrowtail & VG & \twoheadrightarrow & VQ \\
 & \downarrow & & \downarrow & & \downarrow \\
e : & N & \overset{\varkappa}{\rightarrowtail} & G & \overset{\pi}{\twoheadrightarrow} & Q \\
 & \| & & \downarrow{\scriptstyle nat} & & \downarrow{\scriptstyle nat} \\
\psi(e) : & N & \longrightarrow & G/VG & \overset{\pi_*}{\longrightarrow} & Q/VQ .
\end{array}$$

Due to $\varkappa N \cap VG = 0$, the bottom row $\psi(e)$ is an extension. Conversely, given $e_1: N \rightarrowtail E \twoheadrightarrow Q/VQ$ in $\mathfrak{V}$, then $e = (e_1)nat$ satisfies $VG \subseteq VE \times VQ \subseteq 0 \times VQ$ and thus $\varkappa N \cap VG = 0$. Since π maps both VG and $Ker\{G \to E\} \supseteq VG$ isomorphically onto VQ , these subgroups are actually equal. It is now clear that ψ and nat^* are mutually inverse mappings, up to congruence. If e is central, then so is $\psi(e)$. The remaining assertion follows from Theorem I.2.4. A special case of this proposition was already observed by YAMAZAKI [1; Prop. 3.4]. □

The main result of this section is the following

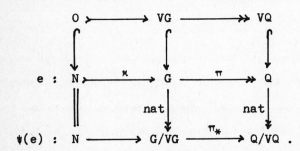

7.19 THEOREM. Let $\mathfrak{V}$ be a variety of groups and $e_i\colon N_i \rightarrowtail G_i \twoheadrightarrow Q_i$ be $\mathfrak{V}$-marginal extensions for $i := 1,2$. Then the following are equivalent:

(i) e_1 and e_2 are $\mathfrak{V}$-isologic extensions;

(ii) there is another $\mathfrak{V}$-marginal extension $e\colon N \rightarrowtail G \twoheadrightarrow Q$ together with $\mathfrak{V}$-isologic epimorphisms $(\sigma_i, \tau_i, \eta_i)\colon e \to e_i$ for $i := 1,2$;

(iii) there exists an isomorphism $\eta\colon Q_1 \to Q_2$ such that $\mathfrak{V}M(\eta)\ \mathrm{Ker}\ \mathfrak{V}\theta_*(e_1) = \mathrm{Ker}\ \mathfrak{V}\theta_*(e_2)$ in $\mathfrak{V}M(Q_2)$.

If G_1 and G_2 are finite, then G in (ii) may be chosen finite, too.

By this theorem and Remarks 7.11, the set of $\mathfrak{V}$-isologism classes of $\mathfrak{V}$-marginal extensions with factor Q is in bijective correspondence with the Aut(Q)-orbits of the Q-subgroups of $\mathfrak{V}M(Q)$. As an immediate corollary, we can also decide whether two groups are $\mathfrak{V}$-isologic or not. LEEDHAM-GREEN/McKAY [1; pp. 113-114] have further results related to (i)$\longleftrightarrow$(iii).

7.20 LEMMA. Let e as in (7.1) be a $\mathfrak{V}$-marginal extension and $\eta\colon Q \to Q_2$ an isomorphism, consider

$$e'\ :\ N \overset{\varkappa}{\rightarrowtail} G \overset{\eta\pi}{\twoheadrightarrow} Q_2 \ .$$

Then $(1,1,\eta^{-1})\colon e' \to e$ and $(1,1,\eta)\colon e \to e'$ are $\mathfrak{V}$-isologic epimorphisms and $\mathrm{Ker}\ \mathfrak{V}\theta_*(e') = \mathfrak{V}M(\eta)\ \mathrm{Ker}\ \mathfrak{V}\theta_*(e)$. Moreover $\tilde{v}_e = \tilde{v}_{e'} \circ \eta^n$ for all n-letter laws $v \in V$.

PROOF. The existence of $(1,1,\eta)$ is obvious and implies $\mathfrak{V}\theta_*(e) = \mathfrak{V}\theta_*(e')\ \mathfrak{V}M(\eta)$ by the naturality assertion of Proposition 7.10. □

This lemma allows one "to pull an isomorphism into the extension" and reduces the proof of Theorem 7.19 to the following characterization of special isologism.

7.21 THEOREM. Given $\mathfrak{B}$-marginal extensions $e_i: N_i \rightarrowtail G_i \twoheadrightarrow Q$ for $i = 1,2$. Then the following are equivalent:

(i) e_1 and e_2 are special $\mathfrak{B}$-isologic extensions;

(ii) there is a $\mathfrak{B}$-marginal extension $e: N \rightarrowtail G \twoheadrightarrow Q$ with $\mathfrak{B}$-isologic epimorphisms $(\sigma_i, \tau_i, 1): e \to e_i$ for $i := 1,2$;

(iii) $\operatorname{Ker} \mathfrak{B}\theta_*(e_1) = \operatorname{Ker} \mathfrak{B}\theta_*(e_2) \subseteq \mathfrak{B}M(Q)$.

If G_1 and G_2 are finite, then G in (ii) may be chosen finite, too.

7.22 LEMMA. Given $\mathfrak{B}$-marginal extensions

$$e_i : N_i \overset{\varkappa_i}{\rightarrowtail} G_i \overset{\pi_i}{\twoheadrightarrow} Q ,$$

consider $e = \langle e_1, e_2 \rangle : N_1 \times N_2 \overset{\varkappa}{\rightarrowtail} G \overset{\pi}{\twoheadrightarrow} Q$ with

$$G = G_1 \curlywedge G_2 = \{ (g_1, g_2) \in G_1 \times G_2 \mid \pi_1 g_1 = \pi_2 g_2 \in Q \}$$

and $\varkappa(n_1, n_2) = (n_1, n_2)$ and $\pi(g_1, g_2) = \pi_1 g_1$. Then e is $\mathfrak{B}$-marginal and

$$\operatorname{Ker} \mathfrak{B}\theta_*(e) = \operatorname{Ker} \mathfrak{B}\theta_*(e_1) \cap \operatorname{Ker} \mathfrak{B}\theta_*(e_2) .$$

PROOF. There are obvious morphisms

$$(7.9) \qquad (\sigma_i, \tau_i, 1) : \langle e_1, e_2 \rangle \to e_i$$

with $\sigma_i(n_1, n_2) = n_i$ and $\tau_i(g_1, g_2) = g_i$. Since words in $G_1 \times G_2$ are evaluated componentwise, cf. (5.3), e is $\mathfrak{B}$-marginal. The last assertion is analogous to Lemma III.2.2, with (7.6) employed instead of I.(3.3'). Indeed (7.9) yields

$$\mathfrak{B}\theta_*(e_i) = \sigma_i \circ \mathfrak{B}\theta_*(\langle e_1, e_2 \rangle) .$$

This and $(\operatorname{Ker} \sigma_1) \cap (\operatorname{Ker} \sigma_2) = 0$ imply the formula for $\operatorname{Ker} \mathfrak{B}\theta_*(e)$. □

7.23 PROOF of Theorem 7.21 (completing the proof of Theorem 7.19). Clearly (ii) implies (i). We now prove (iii) from (ii). Given morphisms $(\sigma_i,\tau_i,1)\colon e \to e_i$ of $\mathfrak{V}$-marginal extensions by Q, Remark 7.12 give commutative diagrams

$$\begin{array}{ccccccccc}
\operatorname{Ker} \mathfrak{V}\theta_*(e) & \hookrightarrow & \mathfrak{V}M(Q) & \xrightarrow{\theta} & VG & \longrightarrow & VQ & \longrightarrow & 0 \\
\downarrow{\scriptstyle \alpha_i} & & \| & & \downarrow{\scriptstyle \beta_i} & & \| & & \\
\operatorname{Ker} \mathfrak{V}\theta_*(e_i) & \hookrightarrow & \mathfrak{V}M(Q) & \xrightarrow{\theta_i} & VG_i & \longrightarrow & VQ & \longrightarrow & 0
\end{array} \tag{7.10}$$

where α_i is an inclusion and β_i the restriction of τ_i. Assuming that $(\sigma_i,\tau_i,1)$ are special $\mathfrak{V}$-isologic epimorphisms, we conclude that β_1 and β_2 are isomorphisms. Consequently,

$$\operatorname{Ker} \mathfrak{V}\theta_*(e_1) = \operatorname{Ker} \theta_1 = \operatorname{Ker} \theta = \operatorname{Ker} \theta_2 = \operatorname{Ker} \mathfrak{V}\theta_*(e_2) .$$

Finally, given $\mathfrak{V}$-marginal extensions e_1 and e_2 by Q as in (i) or (iii). We invoke $e = \langle e_1,e_2\rangle$ and $(\sigma_i,\tau_i,1)\colon e \to e_i$ from Lemma 7.22. Clearly τ_1 and τ_2 are surjective with

$$\tau_1^{-1}(\varkappa_1N_1) = N_1 \times N_2 = \tau_2^{-1}(\varkappa_2N_2) .$$

We claim that τ_1 and τ_2 are special $\mathfrak{V}$-isologic epimorphisms under either assumption. First assume (iii). Then

$$\operatorname{Ker} \mathfrak{V}\theta_*(e_1) = \operatorname{Ker} \mathfrak{V}\theta_*(e) = \operatorname{Ker} \mathfrak{V}\theta_*(e_2)$$

by Lemma 7.22. Thus α_1 is the identity map in (7.10), then β_1 is isomorphic and finally $(\operatorname{Ker} \tau_1) \cap VG = 0$. Likewise τ_2 is a $\mathfrak{V}$-isologic epimorphism. Now assume (i) rather than (iii). We again claim $(\operatorname{Ker} \tau_1) \cap VG = 0$. To this end (cf. the proof of 5.4) every element $z \in VG$ has the form

$$z = v((x_1,y_1),\ldots,(x_n,y_n)) = (v(x_1,\ldots,x_n),v(y_1,\ldots,y_n))$$

with $v \in VF_\infty$, $x_1,\ldots,x_n \in G_1$, $y_1,\ldots,y_n \in G_2$, and $\pi_1x_i = \pi_2y_i$. If $z \in \operatorname{Ker} \tau_1$, then $v(x_1,\ldots,x_n) = 1$ and $\tilde{v}_1(\pi_1x_1,\ldots,\pi_1x_n) = 1$, hence $v(y_1,\ldots,y_n) = \tilde{v}_2(\pi_2y_1,\ldots,\pi_2y_n) = \xi\, \tilde{v}_1(\pi_1x_1,\ldots,\pi_1x_n) = 1$. Thus $(\operatorname{Ker} \tau_1) \cap VG = 0$, by symmetry $(\operatorname{Ker} \tau_2) \cap VG = 0$. □

7.24 EXAMPLE. Let $\mathfrak{B} = \mathfrak{N}_c$, then $\mathfrak{B}$-isologism is the same as "n-isoclinism" in the sense of BIOCH [2] . In view of Example 7.6, two groups G and H are n-isoclinic if there exist isomorphisms $\xi: \Gamma_{c+1}G \to \Gamma_{c+1}H$ and $\eta: G/Z_cG \to H/Z_cH$ such that

$$\tilde{v}_c(\eta g_1,\ldots,\eta g_{c+1}) = \xi\,\tilde{v}_c(g_1,\ldots,g_{c+1}) .$$

By Theorem 7.19, n-isoclinic groups G and H determine another group K and n-isoclinic epimorphisms $\tau_1: K \to G$ and $\tau_2: K \to H$ with

$$(\mathrm{Ker}\ \tau_1) \cap \Gamma_{c+1}K = 0 = (\mathrm{Ker}\ \tau_2) \cap \Gamma_{c+1}K .$$

(This particular result is due to BIOCH [2].) By Theorem 7.19 and Example 7.7, n-isoclinism with common quotient group Q is controlled by certain subgroups of

$$\mathfrak{N}_cM(Q) = \frac{R \cap \Gamma_{c+1}F}{[F,\ldots[F,R]\ldots]} ,$$

in terms of any free presentation of Q . In the case of $c = 2$ and the dihedral group D_8 of order 8, LEEDHAM-GREEN/McKAY [1;p.115] have computed $\mathfrak{N}_2M(D_8) \cong \mathbb{Z}/4 \times \mathbb{Z}/2$ which carries a non-trivial D_8-action.

7.25 EXAMPLE. The reader is warned that the analogue of Proposition III.2.6(a) does not hold: we give an example of a $\mathfrak{B}$-isologism class that does not contain a $\mathfrak{B}$-stem extension; the latter is defined as a $\mathfrak{B}$-marginal extension e such that $\mathfrak{B}\theta_*(e)$ is surjective.

Let $\mathfrak{B}$ be the variety of abelian groups of exponent dividing $n = p^2 \neq 1$, let $Q = \mathbb{Z}/p$. Then $\mathfrak{B}$ is defined by the law $x_1^n[x_2,x_3]$, an extension e as in (7.1) is $\mathfrak{B}$-marginal precisely when e is central and N has exponent dividing n . We start with the obvious free presentation $\bar{e}_p$ as in (2.1) and obtain the $\mathfrak{B}$-marginal extension $m(\bar{e}_p)$ by Q . Let us assume that the $\mathfrak{B}$-isologism class of $m(\bar{e}_p)$ contains a $\mathfrak{B}$-stem extension e . Then

Remark 7.11(a) gives rise to the commutative diagram

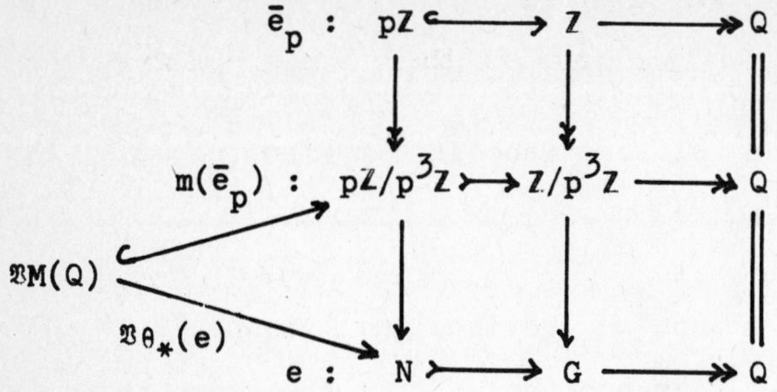

We conclude

$$\mathfrak{B}M(Q) = \mathrm{Ker}\{pZ/p^3Z \to Z/VZ\} = p^2Z/p^3Z \cong Z/p$$

and $\mathrm{Ker}\ \theta_*(m(\bar{e}_p)) = 0$. If e existed as required, then by Theorem 7.19 $\mathfrak{B}\theta_*(e)$ would be isomorphic and thus $\mathfrak{B}M(Q)$ be a direct summand of $pZ/p^3Z \cong Z/p^2$. Contradiction.

7.26 PROPOSITION. a) If G_1 and G_2 are $\mathfrak{B}$-isologic, then the variety generated by G_1 and $\mathfrak{B}$ agrees with that generated by G_2 and $\mathfrak{B}$. In particular, if G_1 and G_2 are isoclinic, then $\mathrm{varo}(G_1) = \mathrm{varo}(G_2)$.

b) If $\mathfrak{B}$ is a variety of exponent 0 and Q is in $\mathfrak{B}$, and if $G_1 \to Q$ and $G_2 \to Q$ are $\mathfrak{B}$-covering groups of Q , then $\mathrm{varo}(G_1) = \mathrm{varo}(G_2)$.

Part (a) and its proof below are due to P.M. NEUMANN [1], part (b) to LEEDHAM-GREEN [1; Thm. 4.5] - the latter for finite groups. If $\mathfrak{B}$ is a variety of exponent 0 , then a $\mathfrak{B}$-<u>covering group</u> of Q is a stem extension e by Q such that $\mathrm{Ker}\ \theta_*(e) = K_{\mathfrak{B}}(Q)$. Note that $G_1, G_2 \in \mathfrak{B}$ by Theorem 6.3.

PROOF. a) Every law v for the variety $\mathfrak{W}$ generated by G_1 and $\mathfrak{B}$ lies in VF_∞ , hence is evaluated in G_2 in the same way as in the $\mathfrak{B}$-isologic group G_1 . Since $\tilde{v}$ is the trivial function on $(G_1/W^*G_1)^n$, the law also holds in G_2 . All told, $G_2 \in \mathfrak{B}$. For the final assertion recall that isoclinism is $\mathfrak{A}$-isologism and that $\mathrm{varo}(G_1)$ is the variety generated by G_1 and $\mathfrak{A}$.

b) The groups G_1 and G_2 are isoclinic by Theorem III.2.3 with $\eta = 1_Q$. □

7.27 COROLLARY. Let G and H be isoclinic groups of exponents m and n , respectively. (Set exponent = 0 if the group contains elements of infinite order.) If m divides n , then H generates the same variety as G and $\mathbb{Z}/n$ do.

PROOF. By Proposition 7.26(b) varo(G) = varo(H) . Let L and R denote the (maximal) sets of laws defining varo(H) and the variety var(H) generated by H , respectively. Then $L = R \cap [F_\infty, F_\infty]$ and $R = F_\infty^n \cdot L$ by Remark 5.6. Likewise, var(G) has the laws $F_\infty^m \cdot L$, and $(F_\infty^m \cdot L) \cap [F_\infty, F_\infty] = L$. Then the variety generated by G and $\mathbb{Z}/n$ has the laws

$$(F_\infty^m \cdot L) \cap ([F_\infty, F_\infty] \cdot F_\infty^n) = ((F_\infty^m \cdot L) \cap [F_\infty, F_\infty]) \cdot F_\infty^n = L \cdot F_\infty^n = R \quad ,$$

the modular law being applicable because of $F_\infty^n \subseteq F_\infty^m$. □

7.28 EXAMPLE, cf. H. NEUMANN [1; p.176]. Let p be an odd prime number and $Q = \mathbb{Z}/p \times \mathbb{Z}/p$. There are two isomorphism types of representation groups of Q , viz. the non-abelian groups of order p^3 . One is of exponent p , H say, and the other is the metacyclic group $G = G(p^2, p, 1+p, 0)$ of exponent p^2 . By Corollary 7.27 for H and G , var(G) is also generated by H and $\mathbb{Z}/p^2$. In particular $\mathrm{var}(H) \subseteq \mathrm{var}(G)$.

See the papers by LEEDHAM-GREEN/McKAY [1] and MOGHADDAM [1] for successful computations of $\mathfrak{B}M(Q)$. More needs to be done!

BIBLIOGRAPHY

J.L. ALPERIN and KUO Tze-Nan:

[1] The exponent and the projective representations of a finite group. Illinois J. Math. 11 (1967), 410-413.

G. BAUMSLAG, E. DYER, and A. HELLER:

[1] The topology of discrete groups. J. Pure Appl. Algebra 16 (1980), 1-47.

R. BAER:

[1] Erweiterungen von Gruppen und ihren Isomorphismen. Math. Z. 38 (1934), 375-416.

[2] Groups with abelian central quotient group. Trans. Amer. Math. Soc. 44 (1938), 357-386.

[3] Groups with preassigned central and central quotient group. Trans. Amer. Math. Soc. 44 (1938), 387-412.

F.R. BEYL:

[1] The classification of metacyclic p-groups, and other applications of homological algebra to group theory. Dissertation (Ph.D.), Cornell University, Ithaca NY 1972, 100 pp. Available from: University Microfilms International, Ann Arbor-London; Order No. 73-6636.

[2] The Schur multiplicator of metacyclic groups. Proc. Amer. Math. Soc. 40 (1973), 413-418.

[3] Abelian groups with a vanishing homology group. J. Pure Appl. Algebra 7 (1976), 175-193.

[4] Commutator properties of extension groups. C. R. Math. Rep. Acad. Sci. Canada 2 (1980), 27-30.

[5] Isoclinisms of group extensions and the Schur multiplicator; pp. 169-185 in: Groups – St Andrews 1981 (ed. by C.M. CAMPBELL and E.F. ROBERTSON), London Math. Soc. Lecture Note Ser. vol. 71, Cambridge et al. 1982.

F.R. BEYL, U. FELGNER, and P. SCHMID:

[1] On groups occurring as center factor groups. J. Algebra 61 (1979), 161-177.

F.R. BEYL and M.R. JONES:

[1] Addendum to "The Schur multiplicator of metacyclic groups". Proc. Amer. Math. Soc. 43 (1974), 251-252.

J.C. BIOCH:

[1] Monomiality of groups. Dissertation (Dr.), Rijksuniversiteit te Leiden, Leiden 1975.

[2] On n-isoclinic groups. Nederl. Akad. Wetensch. Proc. Ser. A 79 = Indag. Math. 38 (1976), 400-407.

J.C. BIOCH and R.W. van der WAALL:

[1] Monomiality and isoclinism of groups. J. Reine Angew. Math. 298 (1978), 74-88.

A. BRANDIS:
[1] Beweis eines Satzes von Alperin und Kuo Tze-Nan. Illinois J. Math. 13 (1969), 275.

J. BUCKLEY:
[1] Automorphism groups of isoclinic p-groups. J. London Math. Soc. (2) 12 (1975), 37-44.

A.H. CLIFFORD:
[1] Representations induced on an invariant subgroup. Ann. of Math. (2) 38 (1937), 533-550.

H.S.M. COXETER:
[1] The binary polyhedral groups, and other generalizations of the quaternion group. Duke Math. J. 7 (1940), 367-379.

R.H. CROWELL and R.H. FOX:
[1] Introduction to Knot Theory. Ginn and Co., Boston et al. 1963. (Fourth corrected printing: Graduate Texts in Math. vol. 57, Springer-Verlag, New York et al. 1977.)

C.W. CURTIS and I. REINER:
[1] Representation Theory of Finite Groups and Associative Algebras. Pure Appl. Math. vol. 11, Interscience, New York-London 1962.

R.K. DENNIS and M.R. STEIN:
[1] The functor K_2: A survey of computations and problems; pp. 243-280 in: H. BASS (ed.), Algebraic K-Theory II (Battelle Institute Conf., Seattle 1972), Lecture Notes in Math. vol. 342, Springer-Verlag, Berlin et al. 1973.

de SÉGUIER, J. > SEGUIER, J. de

P. DU VAL:
[1] Homographies Quaternions and Rotations. Oxford Math. Monographs, Oxford University Press, Oxford 1964.

B. ECKMANN:
[1] Cohomology of groups and transfer. Ann. of Math. (2) 58 (1953), 481-493.

B. ECKMANN, P.J. HILTON, and U. STAMMBACH:
[1] On the homology theory of central group extensions: I - The commutator map and stem extensions. Comment. Math. Helv. 47 (1972), 102-122.
[2] On the Schur multiplicator of a central quotient of a direct product of groups. J. Pure Appl. Algebra 3 (1973), 73-82.

S. EILENBERG and S. MAC LANE:
[1] Cohomology theory in groups (Abstract). Bull. Amer. Math. Soc. 50 (1944), 53.
[2] Cohomology theory in abstract groups, I. Ann. of Math. (2) 48 (1947), 51-78.

D.B.A. EPSTEIN:
[1] Finite presentations of groups and 3-manifolds. Quart. J. Math. Oxford (2) 12 (1961), 205-212.

L. EVENS:
[1] Terminal p-groups. Illinois J. Math. 12 (1968), 682-699.
[2] The Schur multiplier of a semi-direct product. Illinois J. Math. 16 (1972), 166-181.

W. FEIT:
[1] On finite linear groups. J. Algebra 5 (1967), 378-400.

A. FRÖHLICH:
[1] Baer-invariants of algebras. Trans. Amer. Math. Soc. 109 (1962), 221-244.

R. FRUCHT:
[1] Über die Darstellung endlicher Abelscher Gruppen durch Kollineationen. J. Reine Angew. Math. 166 (1931), 16-29.
[2] Zur Darstellung endlicher abelscher Gruppen durch Kollineationen. Math. Z. 63 (1955), 145-155.

T. GANEA:
[1] Homologie et extensions centrales de groupes. C.R. Acad. Sci. Paris 266 (1968), A556-558.

W. GASCHÜTZ:
[1] Endliche Gruppen mit treuen absolut irreduziblen Darstellungen. Math. Nachr. 12 (1955), 253-255.
[2] Zu einem von B.H. und H. Neumann gestellten Problem. Math. Nachr. 14 (1956), 249-252.

D.M. GOLDSCHMIDT:
[1] The Schur multiplier revisited; pp. 179-187 in: Finite Groups (Sapporo-Kyoto, Sept. 1974; ed. by N. IWAHORI). Japan Society for the Promotion of Science, Tokyo 1976.

E.S. GOLOD and I.R. ŠAFAREVIČ:
[1] Izv. Akad. Nauk SSSR 28 (1964), 261-272. English translation: On class field towers. Amer. Math. Soc. Transl. (2) 48 (1965), 91-102.

O.N. GOLOVIN:
[1] Mat. Sb.N.S. 27 (69), 427-454 (1950).English translation: Nilpotent products of groups. Amer. Math. Soc. Transl. (2) 2 (1956), 89-115.
[2] Mat. Sb. N.S. 28 (70), 431-444 (1951). English translation: Metabelian products of groups. Amer. Math. Soc. Trans. (2) 2 (1956), 117-131.

J.A. GREEN:
[1] On the number of automorphisms of a finite group. Proc. Roy. Soc. London Ser. A 237 (1956), 574-581.

R.L. GRIESS, jr.:
[1] Schur multipliers of the known finite simple groups. Bull. Amer. Math. Soc. 78 (1972), 68-71.
[2] Schur multipliers of the known finite simple groups, II; pp.279-282 in: The Santa Cruz Conference on Finite Groups, Proc. Symp. Pure Math. vol 37, Providence 1980.

K.W. GRUENBERG:
[1] Cohomological Topics in Group Theory. Lecture Notes in Math. vol. 143, Springer-Verlag, Berlin et al. 1970.

O. GRÜN:
[1] Beiträge zur Gruppentheorie,I. J. Reine Angew. Math. 174 (1935), 1-14.

W. HAEBICH:
[1] The multiplicator of a regular product of groups. Bull. Austral. Math. Soc. 7 (1972), 279-296.
[2] The multiplicator of a splitting extension. J. Algebra 44 (1977), 420-433.

V.E. HAEFELI-HUBER:
[1] Ein Dualismus als Klassifikationsprinzip in der abstrakten Gruppentheorie. Dissertation (Dr. phil.), Universität Zürich, Zürich 1948, 132 pp.

M. HALL, jr. and J.K. SENIOR:
[1] The Groups of Order $2^n (n \leq 6)$. Mac Millan, New York 1964.

P. HALL:
[1] The classification of prime-power groups. J. Reine Angew. Math. 182 (1940), 130-141.
[2] Verbal and marginal subgroups. J. Reine Angew. Math. 182 (1940), 156-157.
[3] On groups of automorphisms. J. Reine Angew. Math. 182 (1940), 194-204.
[4] On the construction of groups. J. Reine Angew. Math. 182 (1940), 206-214.

M.E. HARRIS:
[1] A universal mapping problem, covering groups and automorphism groups of finite groups. Rocky Mountain J. Math. 7 (1977), 289-295.

H. HASSE:
[1] Zahlentheorie, 3rd edition. Akademie-Verlag, Berlin 1969. English translation: Number Theory. Grundlehren math. Wissenschaften Band 229, Springer-Verlag, Berlin et al. 1980.

P.J. HILTON and U. STAMMBACH:
[1] A Course in Homological Algebra. Graduate Texts in Math. vol. 4, Springer-Verlag, New York et al. 1971 (2nd corrected printing: undated).

G. HOCHSCHILD and J.-P. SERRE:
[1] Cohomology of group extensions. Trans. Amer. Math. Soc. 74 (1953), 110-134.

H. HOPF:
[1] Fundamentalgruppe und zweite Bettische Gruppe. Comment. Math. Helv. 14 (1942), 257-309.
[2] Nachtrag zu der Arbeit "Fundamentalgruppe und zweite Bettische Gruppe". Comment. Math. Helv. 15 (1942), 27-32.

O. HÖLDER
[1] Die Gruppen der Ordnungen p^3, pq^2, pqr, p^4. Math. Ann. 43 (1893), 301-412.

B. HUPPERT:
[1] Endliche Gruppen I. Grundlehren math. Wissenschaften Band 134, Springer-Verlag, Berlin et al. 1967. (Reprint with errata added, 1979).

N. IWAHORI and H. MATSUMOTO:
[1] Several remarks on projective representations of finite groups. J. Fac. Sci. Univ. Tokyo, Sect. I, 10 (1964), 129-146.

D.L. JOHNSON:
[1] Presentations of Groups. London Math. Soc. Lecture Note Ser. vol. 22, Cambridge University Press, Cambridge et al. 1976

D.L. JOHNSON and E.F. ROBERTSON:
[1] Finite groups of deficiency zero; pp. 275-289 in: Homological Group Theory (Durham Symposium, Sept. 1977, ed. by C.T.C. WALL), London Math. Soc. Lecture Note Ser. vol. 36, Cambridge et al. 1979.

K.W. JOHNSON:
[1] Varietal generalizations of Schur multipliers, stem extensions and stem covers. J. Reine Angew. Math. 270 (1974), 169-183.
[2] A computer calculation of homology in varieties of groups. J. London Math. Soc. (2) 8 (1974), 247-252.

M.R. JONES and J. WIEGOLD:
[1] A subgroup theorem for multipliers. J. London Math. Soc. (2) 6 (1973), 738.
[2] Isoclinisms and covering groups. Bull. Austral. Math. Soc. 11 (1974), 71-76.

W. van der KALLEN:
[1] The Schur multipliers of SL(3,Z) and SL(4,Z). Math. Ann. 212 (1973/74), 47-49.

W. van der KALLEN and M.R. STEIN:
[1] On the Schur multipliers of Chevalley groups over commutative rings. Math. Z. 155 (1977), 83-94.

D.M. KAN and W.P. THURTON:
[1] Every connected space has the homology of a $K(\pi,1)$. Topology 15 (1976), 253-258.

A. KERBER:
[1] Representations of Permutation Groups I. Lecture Notes in Math. vol. 240, Springer-Verlag, Berlin et al. 1971.

A. KERBER and M.H. PEEL:
[1] On the decomposition numbers of symmetric and alternating groups. Mitt. Math. Seminar Univ. Giessen no. 91 (1971), 45-81.

M.A. KERVAIRE:
[1] Multiplicateurs de Schur et K-théorie; pp. 212-225 in: A. HAEFLIGER and R. NARASIMHAN (editors), Essays on Topology and Related Topics; Mémoires dédiés à Georges de Rham. Springer-Verlag, Berlin et al. 1970.

S.C. KING:
[1] Quotient and subgroup reduction for isoclinism of groups. Dissertation (Ph.D.), Yale University, New Haven CN 1978, 71 pp. Available from: University Microfilms International, Ann Arbor-London; Order No. 79-16616.

R. KNÖRR:
[1] Noch einmal Schur - Über die Hebbarkeit projektiver Darstellungen. Mitt. Math. Seminar Giessen no. 149 (1981), 1-15.

C.R. LEEDHAM-GREEN:
[1] Homology in varieties of groups, I. Trans. Amer. Math. Soc. 162 (1971), 1-14.

C.R. LEEDHAM-GREEN and S. McKAY:
[1] Baer-invariants, isologism, varietal laws and homology. Acta Math. 137 (1976), 99-150.

W. LEMPKEN:
[1] The Schur multiplier of J_4 is trivial. Arch. Math. (Basel) 30 (1978), 267-270.

F.W. LEVI:
[1] Notes on group-theory IV-VI. J. Indian Math. Soc. N.S. 8 (1944), 78-91.

G. LEWIS:
[1] The integral cohomology rings of groups of order p^3 . Trans. Amer. Math. Soc. 132 (1968), 501-529.

A.S.-T. LUE:
[1] Baer-invariants and extensions relative to a variety. Proc. Cambridge Philos. Soc. 63 (1967), 569-578.

I.D. MACDONALD
[1] On a class of finitely presented groups. Canad. J. Math. 14 (1962), 602-613.

T. MacHENRY:
[1] The tensor product and the 2nd nilpotent product of groups. Math. Z. 73 (1960), 134-145.

S. MAC LANE:
[1] Cohomology theory in abstract groups. III: Operator homomorphisms of kernels. Ann. of Math. (2) 50 (1949), 736-761.
[2] Homology. Grundlehren math. Wissenschaften Band 114, Springer-Verlag, Berlin et al. 1963.

R. MANGOLD:
[1] Beiträge zur Theorie der Darstellungen endlicher Gruppen durch Kollineationen. Mitt. Math. Seminar Giessen, no. 69 (1966), 46 pp.

P. MAZET:
[1] Sur le multiplicateur de Schur du groupe de Matthieu M_{22} . C.R. Acad. Sci. Paris Sér. A 289 (1979), 659-661.

C. MILLER:
[1] The second homology group of a group; relations among commutators. Proc. Amer. Math. Soc. 3 (1952), 588-595.

G.A. MILLER:
[1] Note on a group of isomorphisms. Bull. Amer. Math. Soc. 6 (1900), 337-339.

J. MILNOR:
[1] Introduction to Algebraic K-Theory. Ann. of Math. Studies vol. 72, Princeton 1971.

M.R.R. MOGHADDAM:
[1] The Baer-invariant of a direct product. Arch. Math. (Basel) 33 (1980), 504-511.

A.O. MORRIS:
[1] Projective representations of finite groups, pp. 43-86 in: Proceedings of the Conference on Clifford Algebra, its Generalization and Application (ed. by A. RAMAKRISHNAN). Matscience, Madras 1971. Math. Reviews 48 # 4097.

A.I. MOSKALENKO:
[1] Sibirsk.Math. Ž. 9 (1968), 104-115. English translation: On central extensions of an abelian group by using an abelian group. Siberian Math. J. 9 (1968), 76-86. Math. Reviews 37 # 312.

B.H. NEUMANN:
[1] Some remarks on infinite groups. J. London Math. Soc. 12 (1937), 120-127.
[2] On some finite groups with trivial multiplicator. Publ. Math. Debrecen 4 (1955/6), 190-194.

H. NEUMANN:
[1] Varieties of Groups. Ergebnisse der Math., Neue Folge Band 37, Springer-Verlag, Berlin 1967

P.M. NEUMANN:
[1] Oral communication, June 1977.

H.N. NG:
[1] Degrees of irreducible projective representations of finite groups. J. London Math. Soc. (2) 10 (1975), 379-384.
[2] Faithful irreducible projective representations of metabelian groups. J. Algebra 38 (1976), 8-28.

D.G. NORTHCOTT:
[1] An Introduction to Homological Algebra. Cambridge University Press, Cambridge 1960.

S. NORTON:
[1] The construction of J_4 ; pp.271-277 in: The Santa Cruz Conference on Finite Groups, Proc. Symp. Pure Math. vol. 37, Providence 1980.

H. OPOLKA:
[1] Projective representations of extra-special p-groups. Glasgow Math. J. 19 (1978), 149-152.

H. PAHLINGS:
[1] Beiträge zur Theorie der projektiven Darstellungen endlicher Gruppen. Mitt. Math. Seminar Giessen, no. 77 (1968), 63 pp. Math. Reviews 39 # 2890.
[2] Groups with faithful blocks. Proc. Amer. Math. Soc. 51 (1975), 37-40.

I.B.S. PASSI:
[1] Induced central extensions. J. Algebra 16 (1970), 27-39.

P. PLATH:
[1] Darstellungsgruppen elementar-abelscher p-Gruppen. Diplomarbeit (Dipl.-Math.), Lehrstuhl D für Mathematik,Rheinisch-Westfälische Technische Hochschule Aachen, Aachen 1979.

E.S. RAPAPORT:
[1] Finitely presented groups: the deficiency. J. Algebra 24 (1973), 531-543.

E.W. READ:
[1] On the centre of a representation group. J. London Math. Soc. 16 (1977), 43-50.

K. REIDEMEISTER:
[1] Einführung in die kombinatorische Topologie. Vieweg, Braunschweig 1932.

R. REIMERS:
[1] Die Berechnung der Stammgruppen einer Isoklinismenfamilie. Diplomarbeit (Dipl.-Math.),Lehrstuhl D für Mathematik,Rheinisch-Westfälische Technische Hochschule Aachen, Aachen 1972.

R. REIMERS and J. TAPPE:
[1] Autoclinisms and automorphisms of finite groups. Bull. Austral. Math. Soc. 13 (1975), 439-449.

G. de B. ROBINSON:
[1] Representation Theory of the Symmetric Group. Mathematical Expositions No. 12, University of Toronto Press, Toronto 1961.

J.J. ROTMAN:
[1] The Theory of Groups, an Introduction. Allyn and Bacon Ser. in Advanced Math., Boston 1965. (Quotes refer to this first edition.)

O. SCHREIER:
[1] Über die Erweiterung von Gruppen, I. Monatsh. Math. Physik 34 (1926), 165-180.
[2] Über die Erweiterung von Gruppen, II. Abh. Math. Sem. Univ. Hamburg 4 (1926), 321-346.

I. SCHUR (also written J. SCHUR):
[1] Über die Darstellung der endlichen Gruppen durch gebrochene lineare Substitutionen. J. Reine Angew. Math. 127 (1904), 20-50.
[2] Untersuchungen über die Darstellung der endlichen Gruppen durch gebrochene lineare Substitutionen. J. Reine Angew. Math. 132 (1907), 85-137.
[3] Über die Darstellungen der symmetrischen und alternierenden Gruppen durch gebrochene lineare Substitutionen. J. Reine Angew. Math. 139 (1911), 155-250.

J. de SÉGUIER:
[1] Sur la représentation linéaire homogène des groupes symétriques et alternés. J. Math. Pures Appl. (6) 6 (1910), 387-436; ibid. 7 (1911), 113-121.

H. SEIFERT und W. THRELFALL:
[1] Lehrbuch der Topologie. B.G. Teubner, Leipzig-Berlin 1934. (English translation available: Textbook of Topology. Pure Appl. Math. Ser. vol. 89, Academic Press, New York et al. 1980).

B. SPLETTSTÖSSER:
[1] Über die Gruppen von Automorphismen. Diplomarbeit (Dipl.-Math.), Lehrstuhl D für Mathematik,Rheinisch-Westfälische Technische Hochschule Aachen, Aachen 1976.

J. STALLINGS:
[1] Homology and central series of groups. J. Algebra 2 (1965), 170-181.

U. STAMMBACH:

[1] Anwendungen der Homologietheorie der Gruppen auf Zentralreihen und auf Invarianten von Präsentierungen. Math. Z. 94 (1966), 157-177.

[2] Homological methods in group varieties. Comment. Math. Helv. 45 (1970), 287-298.

[3] Homology in Group Theory. Lecture Notes in Math. vol. 359, Springer-Verlag, Berlin et al. 1973

R. STEINBERG:

[1] Générateurs, relations et revêtements de groupes algébriques; pp. 113-127 in: Colloque sur la théorie des groupes algébriques. (Brussels, June 1962). Centre Belge de Recherches Mathématiques, Louvain; Gauthiers-Villars, Paris 1962

[2] Generators, relations and coverings of algebraic groups, II. J. Algebra 71 (1981), 527-543.

R.G. SWAN:

[1] Minimal resolutions for finite groups. Topology 4 (1965), 193-208.

K.-I. TAHARA:

[1] On the second cohomology groups of semidirect products. Math. Z. 129 (1972), 365-379.

J. TAPPE:

[1] On isoclinic groups. Math. Z. 148 (1976), 147-153.

[2] Irreducible projective representations of finite groups. Manuscripta Math. 22 (1977), 33-45.

[3] Isoklinismen endlicher Gruppen. Habilitationsschrift, Rheinisch-Westfälische Technische Hochschule Aachen, Aachen 1978; 155 pp.

[4] Autoclinisms and automorphisms of finite groups II. Glasgow Math. J. 21 (1980), 205-207.

[5] Some remarks on the homology groups of wreath products. Illinois J. Math. 25 (1981), 246-250.

W. THRELFALL und H. SEIFERT:

[1] Topologische Untersuchung der Diskontinuitätsbereiche endlicher Bewegungsgruppen des dreidimensionalen sphärischen Raumes I. Math. Ann. 104 (1931), 1-70.

van der KALLEN, W. > KALLEN, W. van der

K. VARADARAJAN:

[1] Groups for which Moore spaces $M(\pi,1)$ exist. Ann. of Math. (2) 84 (1966), 368-371.

L.R. VERMANI:

[1] A note on induced central extensions. Bull. Austral. Math. Soc. 20 (1979), 411-420.

J.W. WAMSLEY:

[1] The deficiency of metacyclic groups. Proc. Amer. Math. Soc. 24 (1970), 724-726.

[2] Minimal presentations for finite groups. Bull. London Math. Soc. 5 (1973), 129-144.

R.B. WARFIELD, jr.:

[1] Nilpotent Groups. Lecture Notes in Mathematics vol. 513, Springer -Verlag, Berlin et al. 1976.

P.M. WEICHSEL:
[1] On isoclinism. J. London Math. Soc. 38 (1963), 63-65.

J. WIEGOLD:
[1] Nilpotent products of groups with amalgamation. Publ. Math. Debrecen 6 (1959), 131-168.
[2] The multiplicator of a direct product. Quart. J. Math. Oxford (2) 22 (1971), 103-105.
[3] Some groups with non-trivial multiplicators. Math. Z. 120 (1971), 307-308.
[4] The Schur multiplier: an elementary approach; pp. 137-154 in: Groups – St Andrews 1981 (ed. by C.M. CAMPBELL and E.F. ROBERTSON), London Math. Soc. Lecture Note Ser. vol. 71, Cambridge et al. 1982.

K. YAMAZAKI:
[1] On projective representations and ring extensions of finite groups. J. Fac. Sci. Univ. Tokyo Sect. I, 10 (1964), 147-195.

H. ZASSENHAUS:
[1] The Theory of Groups, second edition. Chelsea Publ. Co., New York 1958.

INDEX OF SPECIAL SYMBOLS

Section	Symbol	Short explanation
	$\Longleftrightarrow$	logical equivalence
	$\rightarrowtail$	monomorphism
	$\twoheadrightarrow$	epimorphism
	$\cong$, $\rightarrowtail\!\!\!\!\twoheadrightarrow$	isomorphism
	$\square$	end of proof resp. lack of proof
	$\mathbb{N}$	natural numbers
	$\mathbb{Z}$	integers
	$\mathbb{Q}$	rational numbers
	$\mathbb{R}$	real numbers
	$\mathbb{C}$	complex numbers
	GF(q)	field with q elements
	Σ	summation sign
	(m,n), gcd(m,n)	(positive) greatest common divisor
	0	group with one element
	D_n	dihedral group of order n
	S_n	symmetric group on n letters
IV.2.3	$G(m,n,r,\lambda)$	metacyclic group
	$\mathbb{Z}/n$	cyclic group of order n
	$0: G \to H$	trivial homomorphism
	$1: G \to G$	identity homomorphism
IV.2.4	m_*	m-th power map
	1	neutral element (of any group)
	x_y	$= xyx^{-1}$

Section	Symbol	Short explanation
	$[x,y]$	$= xyx^{-1}y^{-1}$
	$\|G\|$	cardinality
	$\|G:U\|$	index of subgroup
I.2	$[A,B]$	mixed commutator subgroup
	$Z(G)$	center of the group G
IV.(3.6)	$Z^*(G)$	obstruction to capability in Z(G)
	G_{ab}	commutator quotient group
	Tor(G)	torsion subgroup (when defined)
	$N \rtimes Q$	semidirect product (N normal)
	$G \times H$	direct product of groups
II.4.1	$G * H$	free product of groups
II.4.10	$G \otimes H$	tensor product of groups
I.3.12	$Q \wedge Q$	exterior square
	$\times_{i\in I} G_i$	restricted direct product
	$\prod_{i\in I} G_i$	(unrestricted) direct product
	$\oplus_{i\in I} G_i$	direct sum (of modules)
I.1.9	$Der(Q,A,\varphi)$	derivations
I.1.1	$H(Q,N,\varphi)$	factor sets modulo transformation sets
	Ext(Q,A)	set of (congruence classes of) abelian extensions
	Cext(Q,A)	dto. of central extensions
I.1	$Opext(Q,A,\varphi)$	dto. of Q-extensions of (A,φ)
	$\equiv$	congruent
	$e\gamma$	backward induced extension
	αe	forward induced extension
II.3.1	e/U	factor extension
I.1.8	$e\|P$	extension restricted to subgroup

Section	Symbol	Short explanation
	$e_1 \to e_2$	morphism of extensions
I.(2.2)	$e_1 \times e_2$	product extension
I.1	$\alpha_\bullet$, $\gamma^\bullet$	middle maps in induced extensions
I.2.1	α_{ab}, α_c	homomorphisms induced by α
IV.7.4	α_m	homomorphism induced by α
	A^G	fixed elements
I.3	$M(G)$, $M(\gamma)$	Schur multiplicator
I.2.7	$\theta^*(e,A)$	some connecting homomorphism
I.3.5	$\theta_*(e)$	some connecting homomorphism
IV.(3.5)	$U(e)$	kernel of $\theta_*(e)$
I.4.4	χ_G, $\chi(G)$	Ganea map
IV.5.7	varo(G)	variety of exponent zero

Section	Special extension
I.3.3	$e(Q)\ :\ R_Q \hookrightarrow F_Q \twoheadrightarrow Q$
III.(1.1)	$e_G\ :\ Z(G) \hookrightarrow G \twoheadrightarrow G/Z(G)$
IV.7.13	$e_G\ :\ V^*(G) \hookrightarrow G \twoheadrightarrow G/V^*(G)$
II.(1.2)	$\sigma_V\ :\ K^* \xrightarrow{\delta} GL(V) \xrightarrow{\tau} PGL(V)$
IV.(2.1)	$\bar{e}_n\ :\ n\mathbb{Z} \hookrightarrow \mathbb{Z} \xrightarrow{\nu} \mathbb{Z}/n$

Cross references with Roman numerals refer to that chapter, without Roman numerals to the local chapter. Items of the bibliography are cited by NAME.

SUBJECT INDEX

Vol. 817: L. Gerritzen, M. van der Put, Schottky Groups and Mumford Curves. VIII, 317 pages. 1980.

Vol. 818: S. Montgomery, Fixed Rings of Finite Automorphism Groups of Associative Rings. VII, 126 pages. 1980.

Vol. 819: Global Theory of Dynamical Systems. Proceedings, 1979. Edited by Z. Nitecki and C. Robinson. IX, 499 pages. 1980.

Vol. 820: W. Abikoff, The Real Analytic Theory of Teichmüller Space. VII, 144 pages. 1980.

Vol. 821: Statistique non Paramétrique Asymptotique. Proceedings, 1979. Edited by J.-P. Raoult. VII, 175 pages. 1980.

Vol. 822: Séminaire Pierre Lelong–Henri Skoda, (Analyse) Années 1978/79. Proceedings. Edited by P. Lelong et H. Skoda. VIII, 356 pages, 1980.

Vol. 823: J. Král, Integral Operators in Potential Theory. III, 171 pages. 1980.

Vol. 824: D. Frank Hsu, Cyclic Neofields and Combinatorial Designs. VI, 230 pages. 1980.

Vol. 825: Ring Theory, Antwerp 1980. Proceedings. Edited by F. van Oystaeyen. VII, 209 pages. 1980.

Vol. 826: Ph. G. Ciarlet et P. Rabier, Les Equations de von Kármán. VI, 181 pages. 1980.

Vol. 827: Ordinary and Partial Differential Equations. Proceedings, 1978. Edited by W. N. Everitt. XVI, 271 pages. 1980.

Vol. 828: Probability Theory on Vector Spaces II. Proceedings, 1979. Edited by A. Weron. XIII, 324 pages. 1980.

Vol. 829: Combinatorial Mathematics VII. Proceedings, 1979. Edited by R. W. Robinson et al.. X, 256 pages. 1980.

Vol. 830: J. A. Green, Polynomial Representations of GL_n. VI, 118 pages. 1980.

Vol. 831: Representation Theory I. Proceedings, 1979. Edited by V. Dlab and P. Gabriel. XIV, 373 pages. 1980.

Vol. 832: Representation Theory II. Proceedings, 1979. Edited by V. Dlab and P. Gabriel. XIV, 673 pages. 1980.

Vol. 833: Th. Jeulin, Semi-Martingales et Grossissement d'une Filtration. IX, 142 Seiten. 1980.

Vol. 834: Model Theory of Algebra and Arithmetic. Proceedings, 1979. Edited by L. Pacholski, J. Wierzejewski, and A. J. Wilkie. VI, 410 pages. 1980.

Vol. 835: H. Zieschang, E. Vogt and H.-D. Coldewey, Surfaces and Planar Discontinuous Groups. X, 334 pages. 1980.

Vol. 836: Differential Geometrical Methods in Mathematical Physics. Proceedings, 1979. Edited by P. L. García, A. Pérez-Rendón, and J. M. Souriau. XII, 538 pages. 1980.

Vol. 837: J. Meixner, F. W. Schäfke and G. Wolf, Mathieu Functions and Spheroidal Functions and their Mathematical Foundations Further Studies. VII, 126 pages. 1980.

Vol. 838: Global Differential Geometry and Global Analysis. Proceedings 1979. Edited by D. Ferus et al. XI, 299 pages. 1981.

Vol. 839: Cabal Seminar 77 – 79. Proceedings. Edited by A. S. Kechris, D. A. Martin and Y. N. Moschovakis. V, 274 pages. 1981.

Vol. 840: D. Henry, Geometric Theory of Semilinear Parabolic Equations. IV, 348 pages. 1981.

Vol. 841: A. Haraux, Nonlinear Evolution Equations- Global Behaviour of Solutions. XII, 313 pages. 1981.

Vol. 842: Séminaire Bourbaki vol. 1979/80. Exposés 543–560. IV, 317 pages. 1981.

Vol. 843: Functional Analysis, Holomorphy, and Approximation Theory. Proceedings. Edited by S. Machado. VI, 636 pages. 1981.

Vol. 844: Groupe de Brauer. Proceedings. Edited by M. Kervaire and M. Ojanguren. VII, 274 pages. 1981.

Vol. 845: A. Tannenbaum, Invariance and System Theory: Algebraic and Geometric Aspects. X, 161 pages. 1981.

Vol. 846: Ordinary and Partial Differential Equations, Proceedings. Edited by W. N. Everitt and B. D. Sleeman. XIV, 384 pages. 1981.

Vol. 847: U. Koschorke, Vector Fields and Other Vector Bundle Morphisms – A Singularity Approach. IV, 304 pages. 1981.

Vol. 848: Algebra, Carbondale 1980. Proceedings. Ed. by R. K. Amayo. VI, 298 pages. 1981.

Vol. 849: P. Major, Multiple Wiener-Itô Integrals. VII, 127 pages. 1981.

Vol. 850: Séminaire de Probabilités XV. 1979/80. Avec table générale des exposés de 1966/67 à 1978/79. Edited by J. Azéma and M. Yor. IV, 704 pages. 1981.

Vol. 851: Stochastic Integrals. Proceedings, 1980. Edited by D. Williams. IX, 540 pages. 1981.

Vol. 852: L. Schwartz, Geometry and Probability in Banach Spaces. X, 101 pages. 1981.

Vol. 853: N. Boboc, G. Bucur, A. Cornea, Order and Convexity in Potential Theory: H-Cones. IV, 286 pages. 1981.

Vol. 854: Algebraic K-Theory. Evanston 1980. Proceedings. Edited by E. M. Friedlander and M. R. Stein. V, 517 pages. 1981.

Vol. 855: Semigroups. Proceedings 1978. Edited by H. Jürgensen, M. Petrich and H. J. Weinert. V, 221 pages. 1981.

Vol. 856: R. Lascar, Propagation des Singularités des Solutions d'Equations Pseudo-Différentielles à Caractéristiques de Multiplicités Variables. VIII, 237 pages. 1981.

Vol. 857: M. Miyanishi. Non-complete Algebraic Surfaces. XVIII, 244 pages. 1981.

Vol. 858: E. A. Coddington, H. S. V. de Snoo: Regular Boundary Value Problems Associated with Pairs of Ordinary Differential Expressions. V, 225 pages. 1981.

Vol. 859: Logic Year 1979–80. Proceedings. Edited by M. Lerman, J. Schmerl and R. Soare. VIII, 326 pages. 1981.

Vol. 860: Probability in Banach Spaces III. Proceedings, 1980. Edited by A. Beck. VI, 329 pages. 1981.

Vol. 861: Analytical Methods in Probability Theory. Proceedings 1980. Edited by D. Dugué, E. Lukacs, V. K. Rohatgi. X, 183 pages. 1981.

Vol. 862: Algebraic Geometry. Proceedings 1980. Edited by A. Libgober and P. Wagreich. V, 281 pages. 1981.

Vol. 863: Processus Aléatoires à Deux Indices. Proceedings, 1980. Edited by H. Korezlioglu, G. Mazziotto and J. Szpirglas. V, 274 pages. 1981.

Vol. 864: Complex Analysis and Spectral Theory. Proceedings, 1979/80. Edited by V. P. Havin and N. K. Nikol'skii, VI, 480 pages. 1981.

Vol. 865: R. W. Bruggeman, Fourier Coefficients of Automorphic Forms. III, 201 pages. 1981.

Vol. 866: J.-M. Bismut, Mécanique Aléatoire. XVI, 563 pages. 1981.

Vol. 867: Séminaire d'Algèbre Paul Dubreil et Marie-Paule Malliavin. Proceedings, 1980. Edited by M.-P. Malliavin. V, 476 pages. 1981.

Vol. 868: Surfaces Algébriques. Proceedings 1976–78. Edited by J. Giraud, L. Illusie et M. Raynaud. V, 314 pages. 1981.

Vol. 869: A. V. Zelevinsky, Representations of Finite Classical Groups. IV, 184 pages. 1981.

Vol. 870: Shape Theory and Geometric Topology. Proceedings, 1981. Edited by S. Mardešić and J. Segal. V, 265 pages. 1981.

Vol. 871: Continuous Lattices. Proceedings, 1979. Edited by B. Banaschewski and R.-E. Hoffmann. X, 413 pages. 1981.

Vol. 872: Set Theory and Model Theory. Proceedings, 1979. Edited by R. B. Jensen and A. Prestel. V, 174 pages. 1981.

Vol. 873: Constructive Mathematics, Proceedings, 1980. Edited by F. Richman. VII, 347 pages. 1981.

Vol. 874: Abelian Group Theory. Proceedings, 1981. Edited by R. Göbel and E. Walker. XXI, 447 pages. 1981.

Vol. 875: H. Zieschang, Finite Groups of Mapping Classes of Surfaces. VIII, 340 pages. 1981.

Vol. 876: J. P. Bickel, N. El Karoui and M. Yor. Ecole d'Eté de Probabilités de Saint-Flour IX – 1979. Edited by P. L. Hennequin. XI, 280 pages. 1981.

Vol. 877: J. Erven, B.-J. Falkowski, Low Order Cohomology and Applications. VI, 126 pages. 1981.

Vol. 878: Numerical Solution of Nonlinear Equations. Proceedings, 1980. Edited by E. L. Allgower, K. Glashoff, and H.-O. Peitgen. XIV, 440 pages. 1981.

Vol. 879: V. V. Sazonov, Normal Approximation – Some Recent Advances. VII, 105 pages. 1981.

Vol. 880: Non Commutative Harmonic Analysis and Lie Groups. Proceedings, 1980. Edited by J. Carmona and M. Vergne. IV, 553 pages. 1981.

Vol. 881: R. Lutz, M. Goze, Nonstandard Analysis. XIV, 261 pages. 1981.

Vol. 882: Integral Representations and Applications. Proceedings, 1980. Edited by K. Roggenkamp. XII, 479 pages. 1981.

Vol. 883: Cylindric Set Algebras. By L. Henkin, J. D. Monk, A. Tarski, H. Andréka, and I. Németi. VII, 323 pages. 1981.

Vol. 884: Combinatorial Mathematics VIII. Proceedings, 1980. Edited by K. L. McAvaney. XIII, 359 pages. 1981.

Vol. 885: Combinatorics and Graph Theory. Edited by S. B. Rao. Proceedings, 1980. VII, 500 pages. 1981.

Vol. 886: Fixed Point Theory. Proceedings, 1980. Edited by E. Fadell and G. Fournier. XII, 511 pages. 1981.

Vol. 887: F. van Oystaeyen, A. Verschoren, Non-commutative Algebraic Geometry, VI, 404 pages. 1981.

Vol. 888: Padé Approximation and its Applications. Proceedings, 1980. Edited by M. G. de Bruin and H. van Rossum. VI, 383 pages. 1981.

Vol. 889: J. Bourgain, New Classes of $\mathcal{L}^p$-Spaces. V, 143 pages. 1981.

Vol. 890: Model Theory and Arithmetic. Proceedings, 1979/80. Edited by C. Berline, K. McAloon, and J.-P. Ressayre. VI, 306 pages. 1981.

Vol. 891: Logic Symposia, Hakone, 1979, 1980. Proceedings, 1979, 1980. Edited by G. H. Müller, G. Takeuti, and T. Tugué. XI, 394 pages. 1981.

Vol. 892: H. Cajar, Billingsley Dimension in Probability Spaces. III, 106 pages. 1981.

Vol. 893: Geometries and Groups. Proceedings. Edited by M. Aigner and D. Jungnickel. X, 250 pages. 1981.

Vol. 894: Geometry Symposium. Utrecht 1980, Proceedings. Edited by E. Looijenga, D. Siersma, and F. Takens. V, 153 pages. 1981.

Vol. 895: J.A. Hillman, Alexander Ideals of Links. V, 178 pages. 1981.

Vol. 896: B. Angéniol, Familles de Cycles Algébriques – Schéma de Chow. VI, 140 pages. 1981.

Vol. 897: W. Buchholz, S. Feferman, W. Pohlers, W. Sieg, Iterated Inductive Definitions and Subsystems of Analysis: Recent Proof-Theoretical Studies. V, 383 pages. 1981.

Vol. 898: Dynamical Systems and Turbulence, Warwick, 1980. Proceedings. Edited by D. Rand and L.-S. Young. VI, 390 pages. 1981.

Vol. 899: Analytic Number Theory. Proceedings, 1980. Edited by M.I. Knopp. X, 478 pages. 1981.

Vol. 900: P. Deligne, J. S. Milne, A. Ogus, and K.-Y. Shih, Hodge Cycles, Motives, and Shimura Varieties. V, 414 pages. 1982.

Vol. 901: Séminaire Bourbaki vol. 1980/81 Exposés 561–578. III, 299 pages. 1981.

Vol. 902: F. Dumortier, P.R. Rodrigues, and R. Roussarie, Germs of Diffeomorphisms in the Plane. IV, 197 pages. 1981.

Vol. 903: Representations of Algebras. Proceedings, 1980. Edited by M. Auslander and E. Lluis. XV, 371 pages. 1981.

Vol. 904: K. Donner, Extension of Positive Operators and Korovkin Theorems. XII, 182 pages. 1982.

Vol. 905: Differential Geometric Methods in Mathematical Physics. Proceedings, 1980. Edited by H.-D. Doebner, S.J. Andersson, and H.R. Petry. VI, 309 pages. 1982.

Vol. 906: Séminaire de Théorie du Potentiel, Paris, No. 6. Proceedings. Edité par F. Hirsch et G. Mokobodzki. IV, 328 pages. 1982.

Vol. 907: P. Schenzel, Dualisierende Komplexe in der lokalen Algebra und Buchsbaum-Ringe. VII, 161 Seiten. 1982.

Vol. 908: Harmonic Analysis. Proceedings, 1981. Edited by F. Ricci and G. Weiss. V, 325 pages. 1982.

Vol. 909: Numerical Analysis. Proceedings, 1981. Edited by J.P. Hennart. VII, 247 pages. 1982.

Vol. 910: S.S. Abhyankar, Weighted Expansions for Canonical Desingularization. VII, 236 pages. 1982.

Vol. 911: O.G. Jørsboe, L. Mejlbro, The Carleson-Hunt Theorem on Fourier Series. IV, 123 pages. 1982.

Vol. 912: Numerical Analysis. Proceedings, 1981. Edited by G. A. Watson. XIII, 245 pages. 1982.

Vol. 913: O. Tammi, Extremum Problems for Bounded Univalent Functions II. VI, 168 pages. 1982.

Vol. 914: M. L. Warshauer, The Witt Group of Degree k Maps and Asymmetric Inner Product Spaces. IV, 269 pages. 1982.

Vol. 915: Categorical Aspects of Topology and Analysis. Proceedings, 1981. Edited by B. Banaschewski. XI, 385 pages. 1982.

Vol. 916: K.-U. Grusa, Zweidimensionale, interpolierende Lg-Splines und ihre Anwendungen. VIII, 238 Seiten. 1982.

Vol. 917: Brauer Groups in Ring Theory and Algebraic Geometry. Proceedings, 1981. Edited by F. van Oystaeyen and A. Verschoren. VIII, 300 pages. 1982.

Vol. 918: Z. Semadeni, Schauder Bases in Banach Spaces of Continuous Functions. V, 136 pages. 1982.

Vol. 919: Séminaire Pierre Lelong – Henri Skoda (Analyse) Années 1980/81 et Colloque de Wimereux, Mai 1981. Proceedings. Edité par P. Lelong et H. Skoda. VII, 383 pages. 1982.

Vol. 920: Séminaire de Probabilités XVI, 1980/81. Proceedings. Edité par J. Azéma et M. Yor. V, 622 pages. 1982.

Vol. 921: Séminaire de Probabilités XVI, 1980/81. Supplément: Géométrie Différentielle Stochastique. Proceedings. Edité par J. Azéma et M. Yor. III, 285 pages. 1982.

Vol. 922: B. Dacorogna, Weak Continuity and Weak Lower Semicontinuity of Non-Linear Functionals. V, 120 pages. 1982.

Vol. 923: Functional Analysis in Markov Processes. Proceedings, 1981. Edited by M. Fukushima. V, 307 pages. 1982.

Vol. 924: Séminaire d'Algèbre Paul Dubreil et Marie-Paule Malliavin. Proceedings, 1981. Edité par M.-P. Malliavin. V, 461 pages. 1982.

Vol. 925: The Riemann Problem, Complete Integrability and Arithmetic Applications. Proceedings, 1979-1980. Edited by D. Chudnovsky and G. Chudnovsky. VI, 373 pages. 1982.

Vol. 926: Geometric Techniques in Gauge Theories. Proceedings, 1981. Edited by R. Martini and E.M.de Jager. IX, 219 pages. 1982.